HARVARD EAST ASIAN MONOGRAPHS

87

Studies in the Modernization of
The Republic of Korea: 1945–1975

The Developmental Role
of the Foreign Sector and Aid

126° 127° 128° 129°

North Korea

Kŭmhwa

Sokch'o

East Sea

38°   38°

KYŎNGGI

Ŭijŏngbu

Ch'unch'ŏn

KANGWŎN

Kangnŭng

Mukho
Pukp'yŏng

Samch'ŏk

Inch'ŏn

Seoul

Yŏju

Wŏnju

Chŏngsŏn

Suwŏn

Ansŏng

Ch'ungju

37°   37°

NORTH
CH'UNGCH'ŎNG

Ch'ŏnan

Ch'ŏngju

Ŭmsŏng

SOUTH
CH'UNGCH'ŎNG

NORTH KYŎNGSANG

Taejŏn

Yŏnmu

Changhang

Kimch'ŏn

P'ohang

36°   36°

Kunsan

I-ri

Okku

Chŏnju

Taegu

Kyŏngju

Yellow Sea

NORTH CHŎLLA

Pŏpsŏngp'o

SOUTH KYŎNGSANG

Chinhae

Chinju

Kwangju

Pusan

35°   35°

SOUTH CHŎLLA

Samch'ŏnp'o

Changsŭngp'o

Sunch'ŏn

Mokp'o

Yŏsu

Mijo-ri

Sin'gŭm-ni

34°

128° 129°

THE REPUBLIC OF KOREA

CHEJU ISLAND

Cheju

0 20 40 60 80 100 120 140 160
KILOMETERS

Sŏgwip'o

126° 127°

0 20 40 60 80 100
MILES

Studies in the Modernization of
The Republic of Korea: 1945-1975

The Developmental Role
of the Foreign Sector and Aid

ANNE O. KRUEGER

PUBLISHED BY
COUNCIL ON EAST ASIAN STUDIES
HARVARD UNIVERSITY

Distributed by
Harvard University Press
Cambridge, Massachusetts and London, England
1979

The Council on East Asian Studies at Harvard University publishes a monograph series
and, through the Fairbank Center for East Asian Research, administers research projects
designed to further scholarly understanding of
China, Japan, Korea, Vietnam, Inner Asia, and adjacent areas.

The Harvard Institute for International Development
is Harvard University's center for interdisciplinary research, teaching, and technical assistance
on the problems of modernization in less developed countries.

The Korea Development Institute
is an economic research center, supported in part by the Korean government,
that undertakes studies of the crucial development issues and prospects of Korea.

Library of Congress Cataloging in Publication Data

Krueger, Anne O
The developmental role of the foreign sector and aid.

(Studies in the modernization of the Republic of
Korea, 1945-1975) (Harvard East Asian Monographs ; 87)
Bibliography: p.
Includes index.
1. Korea—Commerce. 2. Korea—Economic conditions
—1945- 3. Economic assistance, American—Korea.
I. Title. II. Series. III. Series: Harvard East
Asian monographs ; 87.
HF3830.5.Z5K78   330.9'519'04   79-4154
ISBN 0-674-20273-2

Foreword

This is one of the studies on the economic and social modernization of Korea undertaken jointly by the Harvard Institute for International Development and the Korea Development Institute. The undertaking has twin objectives; to examine the elements underlying the remarkable growth of the Korean economy and the distribution of the fruits of that growth, together with the associated changes in society and government; and to evaluate the importance of foreign economic assistance, particularly American assistance, in promoting these changes. The rapid rate of growth of the Korean economy, matched in the less developed world (apart from the oil exporters) only by similar rates of growth in the neighboring East Asian economies of Taiwan, Hongkong, and Singapore, has not escaped the notice of economists and other observers. Indeed there has been fairly extensive analysis of the Korean case. This analysis, has been mainly limited to macroeconomic phenomena; to the behavior of monetary, fiscal, and foreign-exchange magnitudes and to the underlying policies affecting these magnitudes. But there are elements other than these that need to be taken into account to explain what has happened. The development of Korean entrepreneurship has been remarkable; Korea has an industrious and disciplined labor force; the contribution of agricultural development both to overall growth and to the distribution of income requires assessment; the level of literacy and the expansion of secondary and higher education have made their mark; and the combination and interdependence of government and private initiative and administration have been remarkably productive. These aspects together with the growth of urban areas, changes in the mortality and fertility of the population and in public health, are the primary objects of study. It is hoped that they will provide the building blocks from which an overall assessment of modernization in Korea can be constructed.

Economic assistance from the United States and, to a lesser extent, from other countries, has made a sizable but as yet unevaluated contribution to Korean development. A desire to have an assessment undertaken of this contribution, with whatever successes or failures have accompanied the U.S. involvement, was one of the motives for these studies, which have been financed in part by the U.S. Agency for International Development and, in part, by the Korea Development Institute. From 1945 to date, U.S. AID has contributed more than $6 billion to the Korean economy. There has also been a substantial fallout from the $7 billion of U.S. military assistance. Most of the economic assistance was contributed during the period before 1965, and most of it was in the form of grants. In later years the amount of economic assistance has declined rapidly and most of it, though concessional, has been in the form of loans. Currently, except for a minor trickle, U.S. economic assistance has ceased. The period of rapid economic growth in Korea has been since 1963, and in Korea, as well as in other countries receiving foreign assistance, it is a commonplace that it is the receiving country that is overwhelmingly responsible for what growth, or absence of growth, takes place. Nevertheless, economic assistance to Korea was exceptionally large, and whatever contribution was in fact made by outsiders needs to be assessed. One of the studies, *The Developmental Role of the Foreign Sector and Aid,* deals with foreign assistance in macroeconomic terms. The contribution of economic assistance to particular sectors is considered in the other studies.

All the studies in this series have involved American and Korean collaboration. For some studies the collaboration has been close; for others less so. All the American participants have spent some time in Korea in the course of their research, and a number of Korean participants have visited the United States. Only a few of the American participants have been able to read and speak Korean and, in consequence, the collaboration of their colleagues in making Korean materials available has been invaluable. This has truly been a joint enterprise.

The printed volumes in this series will include studies on the growth and structural transformation of the Korean economy, the foreign sector and aid, urbanization, rural development, the role of entrepreneurship, population policy and demographic transition, and education. Studies focusing on several other topics—the financial system, the fiscal system, labor economics and industrial relations, health and social development—will eventually be available either in printed or mimeographed form. The project will culminate in a final summary volume on the economic and social development of Korea.

Edward S. Mason

Edward S. Mason
Harvard Institute
for International Development

Mahn Je Kim

Mahn Je Kim
President,
Korea Development Institute

A Note on Romanization

In romanizing Korean, we have used the McCune-Reischauer system and have generally followed the stylistic guidelines set forth by the Library of Congress. In romanizing the names of Koreans in the McCune-Reischauer systems, we have put a hyphen between the two personal names, the second of which has not been capitalized. For the names of historical or political figures, well-known place names, and the trade names of companies, we have tried to follow the most widely used romanization. For works written in Korean, the author's name appears in McCune-Reischauer romanization, sometimes followed by the author's preferred romanization if he or she has published in English. For works by Korean authors in English, the author's name is written as it appears in the original publication, sometimes followed by the author's name in McCune-Reischauer romanization, especially if the author has published in Korean also. In ordering the elements of persons' names, we have adopted a Western sequence—family name first in all alphabetized lists, but last elsewhere. This is a sequence used by some, but by no means all, Koreans who write in English. To avoid confusion, however, we have imposed an arbitrary consistency upon varying practices. Two notable exceptions occur in references to President Park Chung Hee, and Chang Myon, for whom the use of the family name first seems to be established by custom and preference. Commonly recurring Korean words such as si (city) have not been italicized. Korean words in the plural are not followed by the letter "s." Finally, complete information on authors' names or companies' trade names was not always available; in these cases we have simply tried to be as accurate as possible.

Contents

Contents

Tables

Abbreviations

AA	Automatic Approval list (for imports)
BOK	Bank of Korea
BTN	Brussels tariff nomenclature
c.i.f.	cost, insurance, and freight
CRIK	United Nations, Civil Relief in Korea
DLF	Development Loan Fund
DMB	deposit money bank (loans)
ECA	Economic Cooperation Administration (U.S. aid agency in the late 1940s and early 1950s)
EER	effective exchange rate(s)
EPB	Economic Planning Board
ERP	effective rate of protection
FOA	Foreign Operations Administration (a successor to ECA)
f.o.b.	free on board
GARIOA	Government Appropriations for Relief in Occupied Areas
GDCF	gross domestic capital formation
GDP	gross domestic product
GNP	gross national product
IBRD	International Bank for Reconstruction and Development (The World Bank)
ICA	International Cooperation Administration
IMF	International Monetary Fund
KCAC	Korean Civilian Assistance Commission
KDB	Korea Development Bank
KOTRA	Korean Trade Promotion Corporation
LDC	less developed country

MCI	Ministry of Commerce and Industry
MPC	military payments certificates
MSA	Mutual Security Agency
NER	nominal exchange rate(s)
PL 480	Public Law 480 (under which financing for food imports could be obtained)
PLD EER	price-level-deflated effective exchange rate(s)
PPP EER	purchasing-power-parity effective exchange rate(s)
QR	quantitative restrictions(s)
SITC	Standard International Trade Classification
USAID	United States Agency for International Development
USAMGIK	United States Military Government in Korea
UNC	United Nations Command
UNKRA	United Nations Korea Reconstruction Agency
UNRRA	United Nations Relief and Rehabilitation Administration

Preface

This study is part of the Harvard-Korea Development Institute project on the 30-Year Modernization of Korea. When David Cole and Edward Mason approached me to undertake the "Trade and Aid" study, my curiosity about the role of trade and aid in Korea overcame my reluctance to cover ground much of which had already been explored. For anyone interested in the general topics of international trade and economic development, the South Korean experience is of enormous interest. In a world where developing country after developing country had adopted import substitution and exchange control, with the "foreign trade bottleneck" perceived as the main determinant of the growth rate, South Korea's experience has been exceptional. To be sure, Singapore and Hong Kong have relied upon exporting, but it can be argued—not necessarily with justification—that their lack of rural sector makes their experience "different." Taiwan's experience is more similar, but political problems, her considerably smaller size, and the fact that growth has been somewhat less spectacular all make the Korean experience seem the more fascinating. Moreover, Korea's resource base is deplorably poor, and I can recall, from graduate school days, the apparent hopelessness of Korea's development prospects. To try to learn more firsthand about the role of trade and aid in that transformation was too attractive a prospect to pass up.

Reluctance stemmed from the availability of a number of good studies about Korea's trade. Two in particular cover Korea's trade experience well. The first is by Charles R. Frank, Jr., Kwang Suk Kim, and Larry E. Westphal,[1] who were my colleagues in the National Bureau of Economic Research project. The second is by Kim and Westphal.[2] For many aspects of Korea's trade and development, those works are definitive, and I have relied heavily upon them. Their influence should be evident throughout this study.

I finally decided to undertake the project when I learned that I would be able to collaborate with Wontack Hong and Suk Tai Suh in carrying out the research for the project. They have been invaluable collaborators, not only in undertaking companion research papers,[3] but also in pointing out best data sources, answering numerous questions, and discussing many aspects of Korea's development. They also read the manuscript and made numerous comments which have improved both accuracy and content.

I am also indebted to Edward S. Mason and Kwang Suk Kim, who commented upon the entire manuscript. Kwang Suk Kim was an invaluable source of information at all stages of the research. Larry E. Westphal read the penultimate draft of the manuscript and made many valuable comments and suggestions.

Perhaps the greatest debt, however, is to Mahn Je Kim. He extended KDI facilities, including the time of Drs. Hong and Suh for the research, and also was instrumental in enabling me to interview a number of prominent Korean businessmen and government officials. Their patience and courtesy in discussing Korea's modernization are greatly appreciated.

The project was financed by HIID as part of the Korean modernization study. Edward S. Mason provided intellectual leadership. I am indebted to him and to David Cole for both financial support and valuable insights into the Korean economy.

Pat Kaluza, Carol Such, Linda Lee, and Judy Boher typed the manuscript in original and revised forms. My thanks to them for their cheerfulness, even when confronted with marked-up, undecipherable pages.

Anne O. Krueger
June 1978
Minneapolis, Minnesota

Introduction

Among the many transformations that have accompanied Korea's modernization, perhaps none has been more startling than the shift in the role of trade and aid. As late as the end of the 1950s, Korea was a developing country with many of the "typical" problems. The development effort was geared at import substitution; a chronically overvalued exchange rate was maintained through quantitative restrictions upon imports, multiple exchange rates, and related measures. Imports were financed chiefly by aid, as exports—which were predominantly primary commodities—had failed to grow significantly and were under $30 million. By the mid-1970s, the role of trade and aid in Korea was entirely transformed: exports constituted one of the chief "engines of growth" of the economy; export earnings had increased at an average annual rate in excess of 40 percent for more than a decade. Aid had been replaced to a large extent by export earnings and by private commercial capital flows. Exports, which in 1960 had constituted a mere 2 percent of GNP, were over 28 percent of GNP by 1975.

Understanding the role of trade and aid is crucial for interpreting Korea's recent economic history. It is also of considerable importance in terms of the lessons that may be gleaned for other countries' development policies and prospects. Not only was the Korean economy transformed over the period of modernization, but Korea's export performance has been unmatched by any other developing country. The fact that *both* the growth rate of export earnings *and* the growth rate of real GNP accelerated in much the same time interval raises important questions about the relationship between the two changes, and also about those aspects of the experience that were unique to Korea.

No detailed itemization of the "lessons" emerging from Korea's trade and aid will ever be definitive. Too many changes took place simultaneously, and too many complex interrelationships are involved for any precise quantitative estimate of the importance of the trade-and-aid sectors in Korea's performance. It is, nonetheless, the purpose of this study to provide the evidence available on the role of trade and aid in Korea's development, and to analyze, to the extent techniques of economic analysis permit, the contribution of trade and aid to Korea's modernization.

The study is organized chronologically, primarily because the distinct periods of Korea's economic and political history dictate that approach: the various aspects of trade and aid within each period are best understood in relationship to each other. After analysis of those periods, an assessment of the microeconomic and macroeconomic efficiency of trade and aid is attempted in the last two chapters.

Chapter 1 covers the period from 1945 to 1953, beginning with the departure of the Japanese and lasting until the end of the Korean War. American military occupation started in 1945 and continued through 1948. The disruption of economic activity that accompanied the shift from Japanese rule to U.S. Military Government resulted in pressing needs for relief supplies. During the years of military government, aid was devoted to "relief," or maintenance objectives. Exports were negligible and unimportant as a component of domestic economic activity and also as a source of foreign exchange. Imports, by contrast, were sizable but mostly aid-financed. Despite the short-term nature of the objectives during most of the period of military government, certain reforms were accomplished that were important in laying the foundation for future development. These included land redistribution, the disposition of Japanese properties, and the start of a Korean school system to replace the prior Japanese one.

With the end of the military government in 1948, military relief was replaced by aid administered by the Economic Cooperation Administration (ECA). Little more than continuing relief had been achieved by the ECA when the invasion from the north took place. ECA operations were suspended indefinitely, and the military assumed responsibility for relief operations for the duration of the hostilities. By 1950, there had been substantial recovery from the disruptions associated with the departure of the Japanese and with the partition of the country. It is probable that recovery would have continued, and aid—as contrasted with relief supplies—would have assumed sizable proportions in 1950 had not the war broken out. As it was, the next few years were dominated by the war and its effects. Relief imports, again directed at prevention of "disease, starvation and unrest" of the civilian population, were under military control. The United Nations Korea Reconstruction Agency, which had been voted into existence in the fall of 1950, was not permitted to begin reconstruction activities until 1952.

In terms of Korean modernization, therefore, the early period up to 1953 represents a time when the role of aid was primarily that of relief and "buying time." "Trade" consisted primarily of aid- and military-related imports, so that it is the aid story that dominates the period.

Chapter 2 covers the period of recovery, aid dependence, and emphasis upon import substitution which lasted until 1960. Export earnings were a relatively minor source of foreign exchange during that period, since aid financed the bulk of Korea's exports. Korean trade-and-payments policies were geared primarily to receiving as much aid as possible, preventing the excess demand for foreign exchange from being realized, maintaining an overvalued exchange rate, and stimulating domestic economic activity in import-substitution industries. These policies came into sharp conflict with the perceptions of the American government about desirable trade and balance-of-payments policies. Consequently, trade-and-payments policy became a central point of contention in the aid relationship.

The years 1961 to 1965 marked a time of transition, during which policy changes and the start of rapid export growth virtually transformed the economy. Those changes and developments during the transition years are analyzed in Chapter 3. Chapter 4 focuses on the period from 1966 to 1975, which was dominated by a growth strategy aimed at promoting Korean exports, not only in order to enable the financing of needed imports, but also to provide the major "engine" of growth. During this last period, aid diminished and finally became negligible as a source of foreign exchange, although it was replaced in some measure by inflows of private capital, attracted by deliberate policy and the apparent safety assured by the rapid growth of export earnings.

The final two chapters are concerned with analyzing the role of trade, aid, and capital flows in Korea's growth. Chapter 5 is concerned with the microeconomic aspects of the trade-and-payments regime. The focus is on the efficiency of the import substitution drive, aid, the export-promotion policies of the 1960s and early 1970s, and of capital inflows and their allocation. A final chapter then attempts to place trade, aid, and capital flows in perspective in terms of their contribution to the modernization of Korea in the thirty years after 1945.

ONE

The 1945-1953 Period

The years from 1945 to 1953 were marked by severe economic dislocation, associated first with the departure of the Japanese, then with partition, and finally with the Korean War. Understanding the pattern of trade and aid in those years is important for several reasons. First, the disruptions of the period generated the initial conditions for later reconstruction; understanding of the trade-and-aid policies of later years is not possible without knowledge of prior events and of the extreme economic difficulties that prevailed. Second, many of the issues of later periods had their origins in the 1945-1953 period. For example, controversy over exchange-rate policy, a feature of the aid relationship in the 1950s, started during the Korean War years. Third, despite the fact that many of the achievements of the reconstruction period were lost during the Korean War, some accomplishments endured and contributed importantly to later development.

For thirty-five years prior to 1945, Korea had been a Japanese colony, and trade ties had naturally been determined largely by the Japanese Colonial Government and Japanese entrepreneurs who commanded the vast majority of resources in Korea. Examination of the South Korean production pattern prior to World War II yields insights into both the potential comparative advantage of the partitioned country and the immediate post-war structural imbalances which early imports, financed almost exclusively by the military, were designed to remedy. With respect to the immediate post-war years and up to 1953, as would be expected in the context of triple dislocation, data are scattered and those that exist are of questionable reliability. Nonetheless, they serve to give some idea of the quantitative magnitude of the imbalances of the period. During 1946–1948, the U.S. Military Government undertook some fundamental reforms and also provided relief supplies. In 1949, responsibility for administering aid was shifted to the Economic Cooperation Administration (ECA), which continued many of the programs started earlier. Further changes, of course, ensued in the Korean War and post-war years.

PRODUCTION AND TRADE PATTERNS
PRIOR TO 1945

Data on production and trade prior to 1945 are scattered, and those available pertain primarily to the Korean peninsula as a whole. Available information, however, seems sufficient to permit confidence that exports constituted a sizable fraction of production for a number of key commodities.

To turn first to production, agriculture was dominant, with more than 80 percent of the labor force and about half of national income originating in that sector. It is noteworthy that manufacturing apparently constituted a somewhat higher fraction of output than one would have expected for a country at Korea's stage of development: about 31 percent of gross commodity output originated in manufacturing in 1936.[1]

Data on trade flows reflect these same general patterns. Rice was the largest single export, constituting somewhere between 40 and 53 percent of each year's exports between 1930 and 1935. Consonant with the relatively high share of manufacturing output, however, Korea was already exporting manufactures: trade data suggest that textile exports (mostly silk) exceeded 10 percent of total exports after 1923, while pulp and paper, pig iron, sugar, wheat flour, leather, cement, and ammonium sulphate constituted another 3 to 6 percent of the value of exports.[2]

Although Korea appears to have been a net importer of most manufactured products, data indicate that almost the entire output of raw silk was exported, and that as much as about 40 percent of ammonium sulphate, cement, and sugar output, and about 20 percent of cotton fabrics and paper products output was exported in the mid-1930s.[3]

During the colonial period, Korea's trade balance was negative in every year, with imports constituting as much as 35 percent more than exports in some years, and averaging about 28 percent in excess of exports for the 1936–1939 years.[4] This capital inflow represented, in large part, Japanese investment in Korea.

Japan was Korea's major trading partner throughout the colonial period, with over 85 percent of imports originating in Japan proper[5] in each year from 1936–1939 and an even higher fraction of exports going to Japan proper until 1937, after which the figure still remained in excess of 70 percent.[6]

There is no reliable way of estimating the real volume of trade in the pre-war period. Exports in 1939 were recorded to be 1,006.8 million yen, while imports were 1,388.5 million yen.[7] If one takes the wholesale price index and deflates the trade statistics on a 1970 base, and then converts the resulting estimates to dollars at the official exchange rate, the computations would indicate that Korean exports, valued in 1970 prices, were on the order of magnitude of $936 million in 1939, while imports were $1,291 million in that year. As nearly half of rice output, about one-third of fishery output, and more than half of mineral output were exported, those estimates of trade flows are roughly consistent with exports constituting approximately 30 percent of national income and a per capita income of $130–$150 (1970 prices).[8]

Evidence with regard to production and trade patterns for the north and south separately is extremely sparse. It is well known that agricultural output was predominant in the south, and that the north produced most of the minerals and electric power. The pattern of production differed for manufacturing as well. Reflecting the north's more favorable endowment of mineral resources, 95 percent of basic chemical production, 72 percent of other chemicals, 99 percent of basic chemical fertilizer, and 97 percent of iron and steel production originated in the north, which also accounted for over half of paper products, coal products, non-metallic minerals, and steel and metal products production. By contrast, the south accounted for 83 percent of tobacco production, 88 percent of fiber spinning, 85 percent of textile fabrics, 75 percent of transport equipment, and 100 percent of electrical machinery production, although this last sector was of negligible size. Overall, 45 percent of all manufacturing gross output originated in the south and 55 percent originated in the north, but those figures conceal the extent to which individual industries were concentrated in different regions.[9]

TRADE FLOWS, 1945–1949

Korea must have been adversely affected by World War II. Whatever effects there were were dwarfed, however, by contrast with post-war events; as the Japanese left, the U.S. Military Government took command over "occupied territory," and the country was partitioned.

There are no reliable data to indicate the extent of the chaos of the immediate post-war period. The departure of the Japanese resulted in the shutdown of many businesses as entrepreneurs and technicians left; the partition of the country resulted in a severing of trade and financial ties, a process not completed until 1948 when the power supply from the north was finally completely shut off; and an influx of refugees from the north further aggravated the situation. As if those factors were not enough, world trade patterns were themselves in disarray, so that disruption in Korea was only one part of the broader problem of post-war rehabilitation.

Against this background, it is hardly surprising that inflation was rampant: the wholesale price index is estimated to have increased 40-fold between June and August of 1945 and to have quadrupled again from then to the end of 1946.[10] There were three types of international transactions: regular trade, smuggling, and relief supplies provided by the United States.

Of these three, regular trade was undoubtedly the least important in the early post-war years, and it remained miniscule by contrast with the volume of relief and aid imports right through the Korean War. Table 1 provides an estimate of the approximate volume of official trade for 1939 and for 1946 through 1953. Any comparison between 1939 and 1946 is necessarily very crude. However, when it is recalled that the 1946 price level was about forty times higher than that of 1939, it is evident that 1946 trade was a tiny fraction of the pre-war volume. Data for 1945 are, of course, unavailable but would undoubtedly show a comparable pattern, at least for the latter half of the year.

If the data in Table 1 are approximately accurate, they would indicate that commercial and government-financed trade came to a complete halt until 1948. Even if one allows for the fact that the 1939 data include the trade of both the north and the south, and even if the estimates of trade volumes for 1946 and 1947 are off by several hundred percent, it is improbable that the real volume of trade through official channels was as much as 1 percent of its 1939 level in 1947.

TABLE 1   Estimated Trade Flows, Pre- and Post-War

	Exports	*Imports*	*Exports*	*Imports*
	1,000s of currency units[b]		*1,000s of won constant 1947 prices*	
1939	1,006	1,388	n.a.	n.a.
1946	50	160	90	290
1947	1,110	2,890	1,110	2,090
1948	7,200	8,860	4,420	5,440
1949	11,270	14,740	5,060	6,620
1950[c]	32,570[a]	5,210[a]	9,360	1,500
1951	45,910	121,830	2,090	5,550
1952	194,960	704,420	4,270	15,410
1953	398,720	2,237,010	6,700	37,590

Sources: 1939 data from Wontack Hong, "Trade Distortions and Employment," Table B.4. Data in current and constant prices for 1946 to 1953 are from Charles Frank, Kwang Suk Kim, and Larry E. Westphal, *Foreign Trade Regimes and Economic Development,* (National Bureau of Economic Research, New York, 1975), p. 10.

Notes: [a]Recorded private and government trade only. Aid-financed goods are not included.
[b]Yen in 1939, wŏn from 1946 to 1951, and hwan in 1952 and 1953.
[c]Does not include trade through Seoul and Inch'ŏn ports, as the data were lost in the war.

For 1948 and 1949, the apparently rapid increase in the volume of trade reflects primarily the incredibly small base from which it started. From the viewpoint of understanding trade and aid in Korean modernization, the essential point is that regular commercial international trade in the period between the end of World War II and the Korean War was negligible. The "recovery" of exports, which consisted almost entirely of agricultural products and minerals, still left exports trivial by pre-war standards when the Korean War broke out.[11]

The outbreak of the Korean War in 1950 again resulted in the complete disappearance of commercial and government-financed trade, as indicated in Table 1. Indeed, all that needs to be recognized about trade flows during the 1946–1953 period is that commercially financed trade was, for all practical purposes, nonexistent.

RELIEF AND AID PRIOR TO THE KOREAN WAR

Several agencies were involved in providing assistance of one sort or another in the 1945–1953 period. For the convenience of the reader, Table 2 lists the main agencies, their periods of operation, and the total relief supplied.

AMERICAN DOMINANCE AND OBJECTIVES

The U.S. Military Government in Korea (USAMGIK) was the first agency, and began officially on September 9, 1945. It continued holding authority until August 1948. The Republic of Korea was then established and recognized by the General Assembly of the United Nations. The United States transferred its supporting assistance, both economic and military, to the Economic Cooperation Administration (ECA) at that time,[12] and ECA aid to the Republic of Korea lasted until April 1951. Then, the ECA mission was closed down due to the war, and its functions were transferred to the United Nations Korea Reconstruction Agency (UNKRA). Thereafter, a bewildering variety of agencies became involved in administering various forms of assistance. During the period prior to the Korean War, GARIOA (Government Appropriations for Relief in Occupied Areas) assistance amounted to about $500 million, spread over the five years 1945 through 1949. The only other assistance received prior to 1949 was from UNRRA (United Nations Relief and Rehabilitation Administration), and it amounted to less than $1 million, due primarily to opposition from the Soviet Union to United Nations assistance for the Republic of Korea. Assistance administered by USAMGIK therefore dwarfed all other sources of imports by a multiple of several hundred.

During the entire 1945–1950 period, American objectives were never clearly defined, and the indecision resulting therefrom was reflected in aid policy, especially as the occupation period progressed. Originally, American occupation was designed to accomplish three purposes: 1) to establish a free and independent Korea (as had been promised at Cairo and Potsdam); 2) to make Korea strong enough to be a stabilizing factor in Asia; and

TABLE 2 Agencies Involved in Relief and Rehabilitation

Agency	Acronym	Period of Operation	Total Assistance ($ millions)
United States Military Government in Korea (USAMGIK)	GARIOA[f]	1945–1949	502.1
Economic Cooperation Administration[a]	ECA	1949–1951	110.9
United Nations Korea Reconstruction Agency	UNKRA[e]	1950–1955	111.6
United Nations Command, Civil Relief in Korea[b]	CRIK[d]	1950–1956	457.2
International Cooperation Administration[c]	ICA	1953–	5.5

Source: Aid magnitudes are from Bank of Korea as reported in *Economics Statistics Yearbook*, various issues.

Notes: [a]ECA goods were received until 1953, although the amount after 1951 was very small—about $4 million.

[b]CRIK commodities were received until 1956; of total assistance, $59.1 million came after 1953.

[c]ICA assistance started in 1953; ICA became the Mutual Security Agency, which in turn became the U.S. Agency for International Development. ICA and its successor agencies were important after 1953; aid for 1953 only is included in the table.

[d]Technically, the U.N. Security Council Resolution of July 31, 1950 created the United Nations Civil Assistance Command (UNCACK), which later became the Korean Civil Assistance Command (KCAC). The latter, in turn, was administered almost exclusively by the American military through CRIK, which nonetheless administered about $35 million of non-U.S. funds. See Harold Koh, "The Early History of U.S. Economic Assistance to the Republic of Korea, 1955-63," typed, 1975, and also Gene M. Lyons, *Military Policy and Economic Aid: The Korean Case, 1950-1953* (Columbus, 1961).

[e]At the same time as UNKRA was created, the U.N. also created UNCURK, the United Nations Commission for the Unification and Rehabilitation of Korea. It was intended to administer economic development assistance for a united Korea after the war. It was not a factor in aid to South Korea.

[f]GARIOA = Government Appropriations for Relief in Occupied Areas, which funded relief supplies for all occupation areas, as the name implies.

3) to make the country a "showcase of democracy" in Asia. It was not entirely clear, however, whether these objectives pertained to a reunified Korea, or whether, instead, these goals could relate to the south standing alone.[13] The problem came to the forefront only when the urgent relief needs had been met, military occupation ended, and ECA assumed responsibility for Korean aid.

THE MILITARY OCCUPATION

In the first years of GARIOA, the feasibility of the three goals did not raise serious questions, since the immediate and pressing need was for supplies that could provide the population with sustenance and enable the restoration of basic economic functions. The former objective was sought with the provision of aid, while the latter centered upon establishing a functioning Korean government, disposing of Japanese-owned properties, and restoring a functioning educational system.

The commodity import program centered on three basic objectives: 1) prevention of starvation and disease; 2) increasing agricultural output; and 3) the provision of basic consumer goods. It is estimated that more than 90 percent of early aid consisted of imports of commodities in finished form which could be immediately distributed without further processing.[14]

Disposing of Japanese properties was necessary both in order to make resources productive and because the American authorities did not wish to be accused of seizing Japanese properties for their own benefit. A major portion of the administrative capabilities of the early occupation period was therefore devoted to issues associated with the disposition of these properties, and especially of land. This ultimately resulted in a land reform of wide-reaching scope.

Since education had been Japanese during the colonial period, trained teachers, as well as buildings and supplies, were lacking. The military authorities therefore were obliged to attempt the restructuring and reform of the educational system, including emphasis on Koreanization and teacher training.

It is not possible to provide any assessment of the relative importance of each of these thrusts of activity of the military government. All that can be done is to provide an account of each, noting that commodity imports were undoubtedly essential in preventing massive starvation and disease and for stimulating agricultural production. Land and educational reforms were oriented toward longer-run goals that could not have been accomplished had not the south been able to survive the early post-war years without the social unrest that would inevitably, in the absence of relief goods, have accompanied the partitioning of the country.

Commodity Imports

The inflow of imports for relief and reconstruction started shortly after the military government began functioning, and continued throughout the period up to the Korean War. Under USAMGIK most expenditures were apparently for consumer goods—especially food, coal, oil, and textiles. Emphasis upon increasing agricultural production also led to the financing of large imports of fertilizers. By 1948, even under USAMGIK, emphasis had begun to shift somewhat toward the provision of goods that would enable increased production capacity in Korea.[15] The frustration of officials in Korea with the short-term nature of American policy was summarized by E. A. G. Johnson:

> The program of economic development which we could attempt was, to be sure, not as far-reaching as we would have liked. Congressional terms of reference restricted the United States army to short-range assistance to Korea, designed primarily to prevent undernutrition, disease, and unrest, and we were constantly accused of going far beyond these guidelines. Yet it was also the function of USAMGIK to facilitate an orderly transition from a Japanese administration to an independent Korean government. To this end, we found it necessary to deviate somewhat from a strict construction of "government and relief in Occupied Territories". We realized we had to engage in a great deal of economic rehabilitation if the new Korean government was to be able to provide a tolerably

satisfactory administration of its inherited capital plant and other resources.[16]

In the first half of 1949, ECA essentially continued GARIOA-type programs, concentrating on the importation of fertilizer, petroleum, agricultural supplies, and other goods. During that year, it also drew up a proposal for a three-year reconstruction program which was duly submitted to Congress. The program was explicitly development oriented. The stated goals were:

1) To maintain a sufficient quantity of consumer goods and raw materials to prevent excessive hardship, disease, and social unrest, and

2) To lay durable foundations for a Korean economy which, with a rapidly diminishing level of subsidy from the United States, could become a solvent trading partner in the world economy.[17]

The proposed aid program focused on three main areas, and was based on the assumption that a viable South Korea would be an exporter of agricultural commodities. The first priority was to be the development of coal resources, itself necessary to achieve the second objective, the expansion of thermal power generating facilities, which was deemed crucial, given the termination of power supplies from the north. Finally, fertilizer production capability was to be developed in line with the view that agricultural exports were to be increased through growth of output of that sector. The ECA request was for $350 million over a three-year period in the expectation that, at the end of that time, private sources of capital and exports would finance imports.

Among American policy-makers there were substantial doubts as to whether South Korea could ever become self-sufficient in the sense of providing an "acceptable" standard of living to her people without substantial aid inflows.[18] The reconstruction program was delayed in Congress in line with these concerns, and appropriations were stop-gap, covering three-month periods while the bill (H.R. 5330) was being debated. On January 19, 1950, the bill failed by one vote. It was then redrafted and reduced in scope. The redrafted version passed and authorized $100 million for fiscal year 1951.

Korean data indicate that, of the total of $110 million administered by the ECA, $23.8 million were received in 1949, $49.3 million in 1950, and $31.9 million in 1951, with the small residual delivered in 1952 and 1953. It is thus evident that the delay in passing the ECA bill effectively prevented the inauguration of any sustained development program before the outbreak of the Korean War.[19]

It is difficult to judge the contribution of GARIOA- and ECA-financed supplies to Korean reconstruction during the 1945–1950 period. Because there are no national income accounts statistics, there is no meaningful aggregate against which to measure aid-financed imports. Moreover, Korean records of commodity imports do not include items imported under aid programs. Even those data that are available are not necessarily reliable, due largely to the nature of the economic situation that then prevailed.

Some idea of orders of magnitude can nonetheless be gleaned from piecemeal evidence. Table 3 gives data on fertilizer imports and grain production and imports in the 1946–1949 period. Food-grain imports constituted as much as 11 percent of the total grain supply in 1947. Wheat imports were considerably larger than rice imports, however, so that aid-financed imports were more important in terms of total grains than they were in terms of rice production.

In addition to the direct augmentation of the food supply, imports of raw materials and agricultural supplies undoubtedly contributed to the growth of domestic agricultural output. As is apparent from Table 3, fertilizer imports grew rapidly over the 1945–1949 period. Some idea of their contribution to the growth of agricultural production can be derived from studies of the determinants of productivity growth within agriculture. The key work is that of Sung Hwan Ban. According to his estimates,

TABLE 3 Contribution of Imports to Domestic
Grain Supplies, 1946–1950

	Production		Imports		Imports/Total Supply		Fertilizer
	Rice	Total Grain	Rice	Total Grain	Rice	Total Grain	Imports
	(1,000 MT)		*(1,000 MT)*		*(%)*		*(1,000 MT)*
1946	n.a.	n.a.	–	164.4	n.a.	n.a.	171.4
1947	2,155	2,806	39.4	353.9	1.7	11.2	419.1
1948	2,403	3,116	69.9	267.3	2.8	7.9	529.3
1949	2,279	3,209	–	57.0	–	1.7	766.1
1950	2,263	3,162	13.3	44.1	0.5	1.3	76.1

Source: BOK, *Economic Statistics Yearbook, 1958.* Agricultural production data, given in sŏk, were converted to metric tons with a conversion factor of 6.45 sŏk = 1 MT.

labor used in agriculture was approximately unchanged over the 1946–1953 period at about the same level as in the pre-war period. By contrast, the total area devoted to crops and total fixed capital stock had declined slightly. Current inputs, and especially fertilizer, by contrast, increased. He estimates fertilizer inputs in 1949 at 11,250 million (1934) yen, compared to a pre-war high of 10,633 million (1934) yen in 1936 and a 1945 low of 3,513 million (1934) yen.[20] According to his data, total agricultural output grew at an annual compound rate of 2.09 percent from 1945 to 1953, with an annual rate of increase of inputs of 1.50 percent. Input increases, therefore, accounted for about 72 percent of output increases. It seems evident, therefore, that imports and fertilizer and other supplies—the component of inputs which was growing—must be credited with a substantial portion of the increase in agricultural output that took place.

The second piece of evidence with regard to the contribution of aid comes from examination of the commodities financed by aid. Table 4 gives the commodity composition of GARIOA imports over the 1945–1949 period. It indicates that foodstuffs

TABLE 4 Commodity Composition of GARIOA Imports, 1945–1949
($ and %)

Commodity	1945 $000	%	1946 $000	%	1947 $000	%	1948 $000	%	1949[b] $000	%
Foodstuffs	3,604	73.0	21,551	43.5	77,574	44.2	67,698	37.7	4,887	5.2
Agricultural Supplies	—	0.0	6,983	14.1	31,394	17.9	38,609	21.5	43,481	46.9
Unprocessed Materials	—	0.0	113	0.2	3,809	2.2	8,093	4.5	11,844	12.8
Petroleum and Fuel	1,330	27.0	12,224	24.7	14,221	8.1	25,510	14.2	9,711	10.5
Medical Supplies	—	0.0	134	0.3	2,096	1.2	3,321	1.8	2,369	2.6
Clothing and Textiles	—	0.0	1,863	3.7	26,680	15.2	5,627	3.1	—	–
Reconstruction[a]	—	0.0	4,994	10.1	17,696	10.1	26,856	15.0	20,172	21.7
Miscellaneous	—	0.0	1,683	3.4	1,911	1.1	3,878	2.2	239	0.3
TOTAL	4,934		49,945		175,371		179,592		92,703	

Sources: BOK, *Economic Review*, (1955), p. 314 for 1945–1948; and *Monthly Statistical Review*, February 1952 for 1949. The categories listed for 1949 do not correspond precisely to those for 1948. Their allocation to the 1945–1948 classification is indicated in Note b.

Notes: [a]"Reconstruction" includes the following categories: automotive, building materials, chemicals and dyestuffs, communications, educational support, fishing industry supplies, highway construction equipment, marine, mining industry, office supplies, power and light, and railroad.

[b]1949 categories of aid goods, when differently classified, were allocated as follows: fertilizer is the only item in Agricultural Supplies; in Unprocessed Materials are raw cotton, spinning raw materials, crossties, bamboo, lumber and "raw materials and semi-finished products"; and Reconstruction includes: chemicals, hides and skins, pulp and paper, cement, salt, iron and steel, machines and equipment, motor vehicle equipment, transport equipment, and rubber products.

and petroleum were the only commodities to reach Korea under the program in 1945. In 1946 and 1947, foodstuffs still accounted for almost half of total imports, while supplies directly for the agricultural sector were another 14–18 percent of total imports. During those early years, "reconstruction" imports were a very small fraction of the total. By 1948 and 1949, however, foodstuffs were decreasing in both relative and absolute importance, while agricultural supplies and reconstruction materials were increasing. To GARIOA imports in 1949 must be added those financed by the ECA in that year although, of course, the latter were relatively small in magnitude.

Comparison of the total imports under GARIOA for each year from 1945 to 1949 with the data for commercial imports given in Table 1 provides further support for the proposition that commercial imports were relatively unimportant in the 1945–1950 period. While it would be desirable to have data indicating the relative importance of aid in GNP and as a fraction of total supply, figures are simply unavailable. Given the destruction and dislocation that had been experienced by the economy, it seems reasonable to conclude that the aid inflow was extremely important in preventing further deterioration in the situation in 1945–1947 and in permitting reconstruction in 1948–1949.

Land Reform
As already briefly mentioned, the departure of the Japanese left a considerable amount of land (and other property) unowned. For lack of alternative, the occupation government vested these lands, and other alien properties, in itself. It was immediately decreed that the maximum rental paid by tenants should be no more than one-third the annual crop. This created some difficulties, in that no records of earlier production levels (and therefore average yields) were available, and the regulation was apparently not enforced for tenants on land owned by Koreans.[21]

The military government was anxious, for political reasons, to divest itself of the vested lands at the earliest possible date[22] and made a number of efforts in that direction which were technically infeasible. However, the New Korea Company, Ltd., which had been established by USAMGIK to administer the vested lands, gradually acquired the records, experience, and staff with which land distribution could be undertaken. Finally, by a military ordinance dated March 1948, the New Korea Company was abolished, and a National Land Administration was established as the Korean government agency responsible for selling the land. No tenant was to be permitted to purchase land if his holdings would increase to more than two chŏngbo (about five acres), and the price paid to the National Land Administration was to be three times annual production, spread over a 15-year period. This was, therefore, approximately equal to about 20 percent of the crop each year. Within a very short time, 700,000 plots had been sold. By September, some 487,621 acres were sold to 502,072 tenants. This represented over 96 percent of all land which had previously been Japanese-owned.[23]

Apparently corruption was held to a minimum. According to Mitchell, who had headed the New Korea Company:

> The office staffs were worked at top speed for a period of a few weeks; morale was kept high; prizes and bonuses were used to keep up production and maintain a competitive spirit in the organization. Most important, however, is the fact that the very momentum of the operation kept it well out in front of sabotage, opposition, and corruption. For example, some Koreans in the National Land Administration admitted privately, and rather sadly, to this observer that they could have become millionaires if they had had time to organize their relatives and send them around to all prospective land purchasers.[24]

The land distribution carried out under American military occupation was followed by a clause in the Korean constitution which called for land reform on Korean-owned lands. Despite some delays, measures were taken in 1949 so that, prior to the Korean War, most land (including that which had been

Korean-owned) had either been redistributed or had been sold privately by landlords anticipating redistribution.[25]

There is considerable debate about the short-run effects of the land distribution and reform on agricultural production. It is estimated that yields in 1948 were about 30–40 percent below their average levels of 1936–1937.[26] Official statistics indicate that 1949 output of grain was not above its 1948 levels, despite the fact that the weather was apparently favorable in that year.[27] It would thus appear that short-run effects on production could not have been positive and were probably somewhat detrimental. Reasons given include the fact that plots were of small size, but also, and perhaps more important, that landlords had previously supplied intermediate inputs to the tenants, who, once redistribution had been accomplished, had no source of supply for those inputs.

Regardless of the short-term impact, however, the consensus is that the long-run effects were strongly positive, especially in political terms. Gregory Henderson regards the land reform undertaken by USAMGIK as the "best" of its accomplishments, and evaluated its effects as follows:

> Tenancy was reduced to about 33 percent from about 75 percent in 1945. The terms were equitable. Disposal of these lands did much to reduce rural instability, undermine Communist influence, actual or potential, among the peasants, increase their cooperation in the election process, and arouse expectation, later fulfilled, that Korean landlord-held lands would be disposed of similarly.[28]

Cole and Lyman reached a similar conclusion:

> In terms of production, the reform was considered to have been somewhat detrimental, at least in the short run. But psychologically and politically it had very positive effects. Subsequent improvements in farm income, though probably resulting as much or more from other factors of production, were in the farmer's mind often connected with the land reform. Moreover, the reform eliminated the fundamental divisive issue in the countryside. Thereafter, the locus of serious political conflict shifted largely if not entirely to the urban centers. The reform similarly eliminated the last key issue

on which the left wing could have hoped to develop substantial rural support in Korea. Finally, it changed the nature of government requirements in the countryside. The basic socio-political obstacle to rural development had been eliminated, but the problems of low productivity and low income remained. It was clear that further improvement in the rural sector would depend upon substantial and relatively sophisticated technical inputs and economic policy management.[29]

Whereas the program to dispose of vested land appears on the whole to have been successful, efforts of the military government to divest itself of Japanese-owned enterprises were less so. By 1948, little progress had been made, and the properties were transferred to the Republic of Korea. Despite an effort in 1949 to persuade landlords dispossessed in the land reform to purchase the enterprises, almost all enterprises remained in government hands until after the end of the Korean War.

Education
The Japanese had operated the Korean school system as a vehicle for "Japanizing" the Koreans: use of the Korean language in the schools was forbidden, and Korean culture and history were not taught. When the U.S. Military Government replaced the Japanese, many former teachers were among those repatriated to Japan. Korean education had to be reorganized both for nationalistic reasons and because the Japanese departure had left a severe shortage of teachers. The military government set out to attempt to "democratize" the Korean educational system. What this meant in practice is not entirely clear, although it did imply increasing educational opportunities for women and a much higher enrollment rate than had been the case under the Japanese.

The American occupation government of necessity placed primary emphasis on elementary education and on increasing both the supply of teachers and their competence. A few numbers serve to indicate the magnitude of the program and its achievements: from 1945 to 1948, the number of elementary school pupils rose 82 percent and the number of secondary school pupils increased 183 percent; simultaneously the number of available teachers increased 55 percent, 569 percent, and 268 percent at the elementary, middle, and secondary levels, respectively.[30] Koh also cites a figure showing that the rate of literacy of adults in Han'gŭl, the Korean alphabet, increased from 20 percent at the time of Liberation to 71 percent two and a half years later. While these figures are only illustrative of the order of magnitude of the program, it seems clear that education programs undertaken in the 1940s contributed importantly to development potential in later years. Of course, the Korean War resulted in the destruction of a high fraction of available class-rooms and educational materials, and aid in the 1950s had once again to be directed toward the education sector.[31]

AID DURING 1950–1953

By the spring of 1950, it could reasonably be said that some momentum in recovery had been achieved: substantial land reform had been accomplished, and agricultural output was considerably above its 1945 level, though it is doubtful whether the per capita consumption levels of the 1930s had been reattained.[32] Some progress had also been made with respect to relieving the disruptions occasioned by the cutoff of electricity. On the industrial side, individual industries had achieved sizable proportionate increases in output, but power bottlenecks, the fact that many enterprises were still in government hands and managed by bureaucrats with little or no industrial experience, and the poor condition of much capital stock still served as con-straints on industrial output. The magnitude of the constraints was lessening over time, however, and the planned ECA program gave promise of more rapid improvement in the next several years.

Once the Korean War started, however, the economy quickly reverted to much the same condition as in 1945–1946: with production declines due to war-associated dislocation and destruction, efforts to rebuild productive capacity ceased, and aid focused on consumer goods and relief supplies to maintain the transport and communications network. Two aspects of the aid relationship are important for purposes of understanding later developments: on the one hand, there is the identification of the aid donors and their role; on the other, important issues arose in attempting to define the Korean and American contributions to the war. Negotiations over this problem set a precedent for much of the aid relationship that was to continue later into the 1950s.

DONORS AND THEIR ROLE

As already mentioned, with the outbreak of the Korean War, it was decided to terminate ECA operations and to turn them over to the United Nations. This was not, in fact, achieved until 1951, although ECA-financed imports in 1951 were only one-fifth of their 1950 level, and contracts for imports of such commodities as cotton and fertilizer were canceled with the outbreak of the war.

As the Korean War was to be waged under U.N. auspices, it was decided that relief and rehabilitation efforts should be administered by the United Nations. Accordingly, the United Nations Korea Reconstruction Agency (UNKRA)[33] was established late in 1950. By that date, U.N. forces had moved north. It was believed that hostilities had ceased and that reconstruction could begin immediately.

Although both UNKRA and the United Nations Military Command were United Nations endeavors, there was an important difference: the Military Command reported to the United Nations in New York via Tokyo and Washington and the American military headquarters there. In effect, the American army was representing the United Nations in running the war. As such, it had the normal powers that an army has to control entry into the war zone, to take measures to protect information that might aid the enemy, to control logistics, and so on.

For a variety of reasons, the United Nations Command exercised this authority to block UNKRA, which reported directly to the United Nations in New York, from beginning its operations for several years. Because of concern with the logistics of military supplies, security, and related issues (such as the confusion that might result if UNKRA shipments were not coordinated with military shipments into the already congested ports), UNKRA was relatively ineffective until 1953. Through CRIK (see Table 2), the army operated its own relief program, designed to prevent "starvation, disease and unrest" in occupied areas. In effect, the U.N. Command took the view that that part of its necessary military function was the maintenance of orderly conditions in areas where the army was functioning—that is, in the part of Korea under the control of the U.N. Command.

The ambiguous role of UNKRA continued through the period of hostilities and up to the signing of the Armistice. The relationship between the American military authorities and representatives of UNKRA was somewhat strained but, without effective backing in Washington, there was little that UNKRA officials could do.[34] John Lewis, in discussing the role of UNKRA, cites one knowledgeable observer who commented, "UNKRA was *kept out!*—for a long time—then allowed to peek—then allowed to plan—then finally allowed to start a puny program in 1952–53."[35]

The relative importance of UNKRA, contrasted with CRIK, can be seen in Table 5. The phasing out of ECA, which had provided over 80 percent of the $58 million received in 1950, is evident. By 1951, CRIK provided more than half of all assistance; in 1952 and 1953, CRIK was dominant as a source of aid, with UNKRA beginning to play a role. The commodity composition of aid during this period is reported below.

DEFINING AID DURING WARTIME

There is a major issue as to what constitutes aid from an ally or allies to a country upon whose territory a war is being fought. To the extent that the ally physically removed from the

TABLE 5 Aid Received, by Source, 1950-1953
($1,000s)

	ECA	CRIK	UNKRA	Total
1950	49,330	9,376	—	58,706
1951	31,972	74,448	122	106,542
1952	3,824	155,235	1,969	161,028
1953	—	158,787	29,580	188,367
TOTAL	85,126	397,846	31,671	514,643

Source: BOK, *Economic Statistics Yearbook, 1961 and 1964*, as given by Wontack Hong, "Trade Distortions and Employment," text Table 4.2.

fighting contributes commodities toward the maintenance of the population in the war-afflicted country, it is not entirely clear whether that is a part of the ally's contribution to the war or constitutes aid. However, it is measured as aid and will be so treated here. CRIK aid was of precisely this nature.[36]

A more difficult issue arises when troops of the allies are maintained on the soil of the country. In that case, the foreign troops must be enabled to purchase some goods and services from the local market even if most goods are provided by the allied military itself. Such was the case with troops under the U.N. Command during the Korean War. When American and other military forces landed, they required a means of payment. To facilitate this, the Korean government turned over to the U.N. Command an "advance" of a large amount of wŏn with the understanding that terms of repayment in foreign exchange would be negotiated. The Korean government continued to make advances at intervals as currency was required for the foreign troops.

A question of major importance thereupon arose: At what exchange rate should the wŏn advances be reimbursed? On the one hand, the wŏn was undoubtedly overvalued at the time of each advance, and Korean insistence upon full payment at the then official exchange rate was, in an economic sense, unreasonable.[37] On the other hand, the injection of wŏn into the stream of purchasing power at a time when the supply of goods and services was shrinking was unquestionably inflationary, and there can be little doubt that, had the Korean government received foreign exchange against the wŏn advances immediately, the increased flow of importables which would thereby have resulted would have damped inflationary pressures at least to some extent. In fact, the American government adopted a bargaining posture under which it: 1) demanded to repay the "wŏn advances" at a "reasonable" exchange rate at the time of repayment; 2) chastised the Korean government for permitting a high rate of inflation; and 3) as a means of applying pressure, refused to pay anything for the wŏn advances while negotiations were under way as to the appropriate exchange rate for repayment.

The Korean government simultaneously adopted a policy of: 1) maintaining the parity of the currency in order to increase bargaining power and thus increase the real proceeds from the wŏn advances; 2) demanding payment against the wŏn advances at the (overvalued) exchange rates that prevailed when the wŏn were issued; and 3) attempting to limit inflationary pressures only insofar as cajoled into doing so as part of a negotiated settlement of each *tranche* of the outstanding wŏn advances.[38]

American concern with Korean inflation had its origins in the ECA period. The Korean government had signed, as a precondition for ECA assistance, a joint protocol identical to that employed in Europe, under which the Korean government agreed to maintain monetary and fiscal stability as part of its contribution toward reconstruction. In April 1950, American concern with Korean inflation had mounted to a point where the then Secretary of State, Dean Acheson, sent a formal memorandum to the Government of the Republic of Korea, in which it was strongly hinted that the entire ECA program would be reassessed unless measures were taken to control inflation.[39]

Obviously, the outbreak of the war intensified inflationary

pressures enormously, and those pressures persisted throughout the 1950s in varying degrees. It is not the purpose here to investigate the causes of inflation, but it is pertinent to note that issues of trade and exchange-rate policy in Korea were decided against a background of inflation, which implied that the Korean currency was becoming increasingly overvalued in the absence of action to devalue it. In addition to the usual reluctance to devalue, the wŏn advances provided a strong incentive for the government to refuse to alter the exchange rate, as this would weaken its bargaining position with respect to the dollar amounts to be received for wŏn advances.[40] At a time when dollars received against wŏn advances were the major source of foreign exchange, it is hardly surprising that this issue created difficulty. Much of the aid relationship in the late 1950s centered around this issue.

TRADE FLOWS DURING THE KOREAN WAR

The Korean War was a period during which the economy lost ground; much of what had been reconstructed during the 1946–1949 period was destroyed as most of South Korea changed hands in the fighting. Trade flows were therefore important primarily in providing commodities to civilians who otherwise would have suffered even more severely through inadequate food, insufficient clothing, inadequate housing, and disease.

As is apparent from Table 1, commerical imports were below their 1949 level in 1951, and probably also in 1950, although records were lost in the war so that the data are incomplete. Imports began growing rapidly in 1952 and 1953, as payment for the first wŏn advances was received and was used to finance a flow of commercial imports in excess of foreign exchange earnings through exports. To be sure, some of the commodities imported must have been sold to troops stationed in Korea, although military imports were separately recorded.

Exports remained low throughout the war. There are no

Korean estimates of the dollar value of commercial flows for
the period prior to 1953, which makes it very difficult to estimate the relative importance of commercial imports contrasted
with aid-financed imports. One approach is to convert the trade
data from Table 1 at the official exchange rate, and then to
compare them with either the IMF estimates of the dollar value
of total imports (f.o.b.) or with the recorded dollar flow of aid-
financed imports. The results of these computations are
reported in Table 6. The results are indicative only of probable
orders of magnitude, but nonetheless are sufficient to show that
aid was far more important than commercial imports which
were, in any event, financed largely by repayment for wŏn
advances rather than by export earnings.

TABLE 6   Indicators of Relative Importance of Aid
and Commercial Imports, 1950–1953

	Commercial Imports (1)	Total Imports (2)	Aid (3)	Aid as % of Total Imports (4)
	($ millions)			
1950	2.89	n.a.	58.7	n.a.
1951	4.06	171.8	106.5	62
1952	11.74	194.7	161.0	83
1953	37.28	314.0	188.4	60

Sources: Column (1). Data from Table 1 converted at the official exchange rate (1.8
for 1950, 3 for 1951, 6 for 1952 and 1953).
Column (2). IMF, *International Financial Statistics*, (May 1976), line 71 rd.
Imports are f.o.b.
Column (3). Table 5.

To all intents and purposes, therefore, the Korean economy of
the years during the war was highly dependent upon imports
and financed those imports by aid or by wŏn advances: commercial foreign exchange earnings were an almost insignificant
factor.

Reliable national income accounts estimates are not available for the period prior to 1953. For 1953, imports are estimated to have been 12.9 percent of GNP, while exports were about 2 percent of GNP.[41] It is probable that that share was slightly greater than in the preceding years, but, even so, it is apparent that imports constituted a major source of supply of commodities, and a significant damper to the inflationary pressures that existed. Indeed, it is difficult to try to imagine what the rate of inflation might have been in the absence of the import surplus.

As mentioned above, CRIK provided the major source of aid-financed imports in the 1950–1953 period. Table 7 gives the commodity composition of CRIK aid, showing that food and clothing comprised the major portion of CRIK supplies in both 1951 and 1952: items such as agricultural equipment, transport equipment, and construction materials, which would have been

TABLE 7 CRIK Supplies Received, by Principal Commodity ($1,000s)

	1951	1952	1953	1954	1955
Foodstuffs	37,746	45,756	73,974	23,397	8,721
(Rice)	(20,121)	(18,537)	(30,236)	(853)	(2,310)
(Barley)	(6,831)	(12,474)	(17,502)	(5,343)	—
Medical and Sanitation Supplies	6,220	5,592	1,742	1,362	1,035
Fuel	555	8,991	12,985	2,810	—
Construction Materials	4,496	5,560	13,260	1,674	2,893
Transportation Equipment	1,947	1,454	347	485	393
Agricultural Equipment	—	23,495	19,874	13,904	14
Rubber and Products	1,039	3,875	709	—	—
Textiles and Clothing	25,444	47,004	33,286	5,037	583
Miscellaneous	—	13,805	2,610	1,472	395
TOTAL	77,447	155,532	158,787	50,141	14,034

Source: BOK, *Economic Review,* various issues.

destined to increase productive capacity, were almost negligible until 1953. Even in that year, food and textiles accounted for more than two-thirds of total supplies provided through CRIK. Thus, if ever aid could be classified as destined for "maintenance," it was the aid provided to Korea during the period of active fighting. Such aid was obviously essential if a higher incidence of malnutrition, exposure, and disease was to be prevented. It did not, however, contribute directly to the modernization of Korea, except in the sense of buying time until reconstruction could begin once again.

THE TRADE-AND-PAYMENTS REGIMES, 1945–1953

Inflation was rampant during the entire 1945–1953 period, although the rate varied from a low of around 60 percent from 1948–1949 and 1952–1953 to a high of 700 percent during 1945. This reflected, of course, severe demand pressures resulting both from the disruption of supply and from the attempts of the government to command resources.

In such circumstances, pressure upon the balance of payments was inevitable under any sort of trade-and-payments regime. The fact that export capacity had been virtually destroyed implied that, without other sources of foreign exchange, imports would be constrained to very small amounts. In fact, there was a strong tendency to attempt further to restrict imports by quantitative techniques while resisting pressures to devalue the currency. The overvaluation of the currency, in turn, implied that withholding foreign exchange earnings, such as they were, from official channels could be extremely profitable if such currency were sold in the black market or used to finance the purchase of goods that could then be imported extra-legally and sold at an extremely high premium. Currency overvaluation also undoubtedly discouraged exports, thereby further intensifying pressures on the balance of payments.

The 1945–1953 period, therefore, was one during which changes in the exchange rate, although frequent, came too little and too late. Simultaneously, exchange-control regulations were continuously altered in an effort to try to channel foreign-exchange transactions through legal mechanisms.

EXCHANGE RATES

Table 8 provides data on the nominal and real exchange rates that prevailed for some major categories of transactions over the period. In the period right after the American occupation began, exchange rates were meaningless, as virtually all private trade was conducted under barter arrangements. Likewise, special export incentives, tariffs on imports, and special exchange rates applicable to them meant that the effective rates for each category of transaction differed, and most were generally above those shown in Table 8.

Nonetheless, the rates given in the two left-hand columns of Table 8 are significant because they reflect the official rate which served as a basis both for government purchases of foreign goods and services, and for remuneration of the wŏn advances. Comparison of those two rates with the U.S. greenback rate (black market rate for dollars) provides one partial indication of the extent to which both the official rate and the greenback rate were overvalued.

Inspection of the exchange rates in Table 8 indicates why the exchange rate could not itself be significant for most private transactions: despite frequent devaluations, the real exchange rate generally failed to keep pace with inflation. In that circumstance, the exchange rate could not serve as a major deterrent to imports or as an incentive for exports. This can be seen in two ways: examination of the behavior of the black market rate, and analysis of changes in the real exchange rate.

Black market rates are not always a reliable indicator of the extent of excess demand for foreign exchange at the prevailing exchange rate. This is especially so when the black market rate is a currency rate, applicable only to cash. For, in that instance, it usually proves possible to use currency only for relatively small transactions, often consumer goods. There is less reason than usual to doubt the validity of the U.S. greenback rate, however, since the rates on U.S. greenbacks and MPC (military payments certificates) applied to (illegal) purchases of goods from the American PX. Even so, the black market rate relative to the official rate must be viewed with considerable caution as an indicator of the extent to which excess demand prevailed.

It is unfortunate that there are no black market quotations prior to 1948. The first one available, for October 1948, indicates a black market rate about 1.7 times the official rate. By June 1949, the gap had widened considerably, with the black market rate equal to about 4.8 times the official rate. The gap between the two rates then narrowed, as the official rate was doubled in May 1950. At the outbreak of the Korean War, the differential between the two rates had fallen to its lowest point, at about 130 percent of the official rate. It stayed within a small range of that differential until December, but thereafter the gap opened markedly. It was during this period that controversy surrounding the exchange rate at which wŏn advances should be repaid was at its height. The differential between the official rate and the black market rate remained huge up to and including the time of the Armistice in August 1953. As can be seen, the black market rate was then about 4.4 times the official rate. While it is difficult to attach very much significance to the precise magnitude of the differential or dates of turning points, premiums as large as 200 and 300 percent are symptomatic of very severe currency overvaluation. While the effective exchange rates[42] for most transactions were usually somewhat above the official rate, they were by no means sufficient to absorb the differential indicated by the data in Table 8.

That overvaluation was severe can also be seen in the right-hand half of Table 8. There, the nominal exchange rates indicated on the left were divided by the wholesale price index (on a 1965 base) to provide an index of the "real" exchange rate. A higher number implies a less overvalued exchange

TABLE 8 Nominal and Effective Exchange Rates, 1945–1953

Effective Date	Nominal Exchange Rates (current wŏn per dollar)			Price Level Deflated Exchange Rates (1965 wŏn per dollar)		
	Official Rate[a]	Counterpart Deposit Rate[b]	U.S. Greenbacks[c]	Official Rate[a]	Counterpart Deposit Rate[b]	U.S. Greenbacks[c]
Oct. 1, 1945	0.015	—	—	93.8	—	—
July 15, 1947	0.05	—	—	29.2	—	—
Oct. 1, 1948	0.44	—	0.74	148.1	—	249.2
Dec. 15, 1948	0.45	0.45	n.a.	137.6	137.6	n.a.
June 14, 1949	0.9	0.45	2.17	252.1	126.1	607.8
Nov. 1, 1949	0.9	0.5	2.55	189.4	105.3	536.8
Dec. 1, 1949	0.9	0.6	2.83	176.1	117.4	553.8
Jan. 1, 1950	0.9	0.8	3.48	159.3	141.6	615.9
Apr. 1, 1950	0.9	0.9	2.98	150.8	150.8	499.2
May 1, 1950	1.8	1.1	2.28	304.1	185.8	385.1
May 15, 1950	1.6	1.1	2.28	270.3	185.8	385.1
June 10, 1950	1.6	1.4	2.42	260.2	227.6	393.5
June 25, 1950	1.8	1.8	2.42	292.7	292.7	393.5
Oct. 1, 1950	1.8	2.5	2.58	162.2	225.5	232.4
Nov. 1, 1950	2.5	2.5	3.42	196.9	196.9	269.3
Dec. 1, 1950	2.5	4.0	6.12	170.1	272.1	416.3

TABLE 8 (continued)

Effective Date	Nominal Exchange Rates (current wŏn per dollar)			Price Level Deflated Exchange Rates (1965 wŏn per dollar)		
	Official Rate[a]	Counterpart Deposit Rate[b]	U.S. Greenbacks[c]	Official Rate[a]	Counterpart Deposit Rate[b]	U.S. Greenbacks[c]
May 1, 1951	2.5	6.0	9.83	91.2	219.0	358.8
Nov. 10, 1951	6.0	6.0	18.21	132.7	132.7	402.9
Average 1952	6.0	6.0	n.a.	71.3	71.3	n.a.
Aug. 28, 1953	6.0	18.0	26.4	55.6	166.7	244.4
Dec. 15, 1953	18.0	18.0	38.7	152.5	152.5	328.0

Source: Frank, Kim, and Westphal, *Foreign Trade Regimes*, pp. 30–32.

Notes: [a]Other rates altered between the dates shown, but the official and counterpart deposit rates did not.
[b]The Counterpart Deposit Rates came into effect with the signing of the ECA-ROK Agreement. U.N. currency advances were subject to slightly different rates for part of the period.
[c]The U.S. greenback rate is the estimated black market rate for U.S. currency.

rate, that is, a higher price of foreign commodities relative to domestic commodities.[43]

The data in Table 8 indicate that the real nominal exchange rate fell sharply after 1945.[44] Even a devaluation of massive proportions in July 1947 failed to restore the real exchange rate to more than one-third its October 1945 level. Successive devaluations in 1948 and 1949 increased the real price of foreign exchange somewhat, and it appears to have reached its highest level in May of 1950.

The impact of the Korean War and the accompanying inflation on the real exchange rate can easily be seen by examining the data for 1950–1953. By August 1953, the real exchange rate was less than 20 percent of its pre-war peak. The extent to which that represented severe currency overvaluation is evident when it is recalled that even the real exchange rate of May 1950 was probably overvalued, although by a smaller proportion than at any other time during the 1945–1953 period. Even the huge devaluation—by 300 percent—of December 1953 failed to restore more than half the erosion in the real rate that had preceded it. This is a factor of considerable importance in interpreting the behavior of exports over the years immediately after the Korean War.

The remaining point to be noted from Table 8 is the relationship between the counterpart-deposit rate and the official rate. The rate to be paid on wŏn advances, the UNC rate, was generally close to the counterpart-deposit rate after 1950, being 2.5 wŏn per dollar from October 1, 1950 to March 11, 1951, 6 wŏn per dollar from then until August 28, 1953, and 18 wŏn per dollar thereafter. Under the original agreement between the United States and the Korean government, wŏn advances were supposed to be repaid in dollars at the official exchange rate prevailing at the time the advances were made by the Bank of Korea. There were very few redemptions at that rate, however, because of the American belief that the government was deliberately refusing to devalue. It was not until February 1953 that an agreement was finally reached under which advances would be redeemed within twenty days, and all outstanding advances from earlier dates were repaid at the rate of 18 wŏn per dollar.[45]

EXCHANGE CONTROL

In the context of rapid inflation, a fixed exchange rate with infrequent and usually insufficiently large devaluations, and virtually nonexistent foreign exchange earnings, it was inevitable that a variety of quantitative restrictions and other measures would have to be taken to absorb excess demand. When, in addition, the Korean government itself received title to foreign exchange, as it did when wŏn advance redemptions became important, the situation was further confounded. From 1945 until 1960 (well past the period under review here) the foreign trade regime became increasingly complex and cumbersome. Most aspects, however, had their origins in the 1946–1953 period, and those developments are worth noting.

First, with regard to exchange control itself, the bulk of trade was initially barter and the Chosŏn Exchange Bank was formed for purposes of administering it in 1947. The period also saw the introduction of tariffs in 1946 and the adoption in 1949 of a tariff system which lasted without major changes until 1957. In addition, the mechanism for licensing of imports and exports was developed in 1946, and the system that prevailed until 1960 was adopted in 1949. Finally, the first export incentives, of a kind that would prevail for much of the rest of the 1950s, were introduced in 1951.

To understand the background against which exchange-control procedures were formulated, it is necessary to recognize that a sizable portion of private foreign trade was conducted under barter arrangements from 1946 to 1953.[46] Initially, in 1946 and 1947, foreign traders, mainly from Hong Kong and Macao, carried out barter trade at southern Korean ports. Koreans were involved primarily as brokers for the Chinese, since they themselves had little experience or capital.

After the Chosŏn Exchange Bank was established, a series of devices was developed to expedite barter trade. One of the early ones was "trust shipping" wherein export documents could, in effect, be accompanied by an order for commodities to be imported.[47] This system was obviously highly inefficient but, nonetheless, provided a sizable incentive to exporters. It set a pattern which has prevailed through much of Korea's modernization: exporting and importing have, to a large extent, been done by the same people.

The years 1948 and 1949 saw a number of measures introduced designed to reduce the restrictiveness of the barter system. In particular, an account was opened with Japan so that exporters were no longer constrained to import through the same Japanese trader to whom they exported. By late 1949, arrangements were being made whereby some expenditure of yen in other countries could also be permitted. These arrangements had not progressed very far, however, when the Korean War broke out. The goods on the docks at Pusan were shipped to Japan on consignment, and other export activity virtually ceased.

Before the outbreak of the Korean War, the Bank of Korea replaced the Chosŏn Exchange Bank, absorbing its exchange-control functions. All private foreign exchange was required by law to be turned over to the Bank of Korea, which held accounts denominated in foreign currencies to guard against exchange risk. Throughout the Korean War, the Bank of Korea was engaged in receiving foreign exchange, in redeeming wŏn advances, and also in administering exchange control. "Loan funds" were established in 1952, which, in effect, required "guarantee deposits" from importers applying for foreign exchange.[48] The system was discontinued in 1954 but was replaced by an equally complex series of charges.

Tariff policy was also crystallizing in the period prior to the Korean War. In 1946, the U.S. Military Government imposed an across-the-board 10 percent tariff, and only commodity imports financed by foreign assistance were exempted. That tariff was replaced in 1949. It is estimated that the new average tariff rate was 40 percent.[49]

The general purposes of the 1949 tariff reform were to raise additional revenue for the government and to give increased protection to goods manufactured in Korea. Thus, food-grains and non-competing capital goods and raw materials were to be imported duty free; 10 percent tariffs were established on "essential" goods for which there was little domestic production, while tariffs of 10 and 20 percent were set respectively for unfinished goods not produced and goods produced in Korea; a comparable distinction was made for finished goods, with rates of 30 and 40 percent, while semi-luxury goods were dutiable at rates ranging from 50 to 90 percent, and luxury goods were taxed at rates of 100 percent and more.

The system increased in complexity over time. The first change was in 1952, when tariff exemptions were set for machinery and equipment imports destined for "key" industries. Despite increasing complexity, however, the basic rate structure introduced in 1949 remained in effect throughout the 1950s.

Along with exchange control and the tariff system, the system of administering quantitative restrictions started in 1946. Import and export licensing was instituted in that year and has prevailed continuously since, although there have been important variations in the nature of the licensing system and the extent to which it has been restrictive.

From 1946 until February 1949, the licensing system was remarkably simple: the authorities announced which commodities would be eligible for importation and which would not; no effort was made to regulate the quantity of eligible imports that would be permitted. In 1949, however, the system was replaced with one that lasted throughout the 1950s, under which import programs (which were quarterly until 1953) were announced, indicating not only the commodities that would be eligible for importation but the amount that could be imported. Under this system, the Ministry of Commerce and Industry licensed imports; quotas were separate for Japan and for other countries.

An indication of the restrictiveness of the import licensing system is provided by the fact that the first major export incentive was an "export-import link" system introduced in 1951.[50] Under this system, certain commodities were designated that might be imported only by exporters. These exporters were then given permission to spend a specified fraction of their foreign exchange receipts—initially 1 to 5 percent, depending on the commodity—on these commodities. To provide an incentive for exporting, the markup on these items must have been substantial. The export-import link system, in one form or another, persisted throughout most of the 1950s and into the 1960s.

As of 1953, therefore, the basic characteristics of the trade-and-payments regime as it would persist throughout the 1950s were already set. The Korean currency was substantially overvalued. Exports were a small fraction of imports which were financed primarily with aid receipts. Imports were permitted only under licensing and subject to quarterly import programs. And import licenses were extremely valuable with the domestic prices of most import commodities well above the amount paid for them by importers.

TWO

Trade and Aid, 1953 to 1960

For the Korean economy, the years from 1953 to 1957 were a period of reconstruction, while the 1958 to 1960 period was one of very slow growth, if not stagnation. For purposes of analyzing trade and aid, however, all these years can be viewed as one period: policy with respect to trade, aid, and exchange rates exhibited a high degree of continuity.

During 1953–1957, economic growth proceeded at a fairly satisfactory rate—about 5 percent per annum in real terms—as the economy recovered from the devastation and destruction of the Korean War. It was also at this time that aid reached its peak, both in absolute terms and as a proportion of resources available to the Korean economy. Although UNKRA provided a significant flow of commodities to Korea in the 1953–1955 period, the United States decided fairly shortly after the end of the war that aid to Korea should be bilateral. Therefore, most of the aid received during the entire period came directly from the United States via bilateral channels, and even that part originating from UNKRA was financed largely by the American contribution.

The trade-and-payments regime was heavily oriented toward import substitution over the entire 1953–1960 period. A major difference between the 1953–1957 period and the 1958–1960 period was that aid flows, which permitted the growth of imports even with stagnant export earnings during the reconstruction years, peaked in 1957 and began declining thereafter. The attempt to continue import substitution after 1957 therefore took place against the background of declining imports, a fact that had significant implications for resource allocation.

Economists learn early in their careers that everything depends on everything else. Seldom has this been more the case than with the various components of trade, aid, and the payments regime during the post-war years. The exchange rate continued to be the football of the aid relationship and did not play a major role in equating the supply and demand for foreign exchange. Quantitative restrictions, therefore, played an important role in allocating scarce foreign exchange. Production for export was discouraged both by the attraction of new resources to the sheltered domestic market and by the relatively unattractive real exchange rates. Imports, therefore, were financed largely by aid funds, while private capital flows, like exports, were negligible. Even with a huge aid-financed import surplus, however, a sizable premium accrued to those who were able to buy imports, as the domestic price of most importables rose well above landed cost. This, in turn, provided a sizable incentive for import-substitution oriented production and also created a source of corruption.

Because of the high degree of interdependence among the various facets of aid, trade, and the payments regime, it is difficult to consider each aspect separately. Nonetheless, it seems desirable to begin by considering the trade-and-payments regime within which exports, imports, and aid functioned, even though it should be recalled that the trade-and-payments regime was partially determined, both directly and indirectly, by negotiations over aid flows. Brief consideration can then be given to the behavior of exports over the reconstruction years. Then some indication of the scope of import substitution can be provided, and the composition of imports and the volume of aid analyzed.[1] Finally, the aid relationship that surrounded exchange-rate negotiations and determination of commodity flows can be discussed.

EXCHANGE RATES
AND EXCHANGE CONTROLS

At the end of the Korean War, a massive 300 percent devaluation was agreed upon.[2] Already there were several different exchange rates in effect for various categories of transactions, and the number of rates, as also the complexity of the rate structure, had increased over the war years. The exchange control and import licensing system that had begun in 1949, however, also continued, with import programs converted to a semi-annual rather than a quarterly basis. The Korean trade-and-payments regime during the 1953–1960 period can, therefore, be best described as a combination of a multiple exchange-rate system and of quantitative controls. Throughout the period, the complexity of the regime was increasing as export incentives, special treatment for commodities imported for specified purposes, and other exigencies led the authorities to devise new regulations, incentives, and categories.

Despite the ever-increasing complexity of the system, import licenses became more valuable as time progressed, at least until 1957. The allocation of aid funds and of licenses became troublesome, as the licenses became increasingly valuable.

EXCHANGE RATES
Nominal rates

Table 9 gives data on nominal and real exchange rates for the 1954–1957 period. The data are comparable with those in Table 8. That is, they represent nominal rather than effective

TABLE 9 Nominal and Real Exchange Rates, 1954–1957

	Official Rate	Counter-part Deposit Rate	Japan Export Dollars	Other Export Dollars	U.S. Greenbacks	Military Payments Certificates (MPC)
			Nominal Rates (current wŏn per dollar)			
Dec. 15, 1953	18.0	18.0	–	–	38.7	29.3
Nov. 10, 1954	18.0	18.0	77.7	74.0	65.6	53.0
Dec. 13, 1954	18.0	18.0	80.9	78.0	71.1	57.6
Jan. 10, 1955	18.0	35.0	92.3	83.5	77.2	62.9
April 18, 1955	18.0	35.0	75.6	46.6	74.8	60.5
Aug. 15, 1955	50.0	50.0	95.0	82.0	80.2	66.2
Average 1956	50.0	50.0	107.0	100.8	96.6	81.0
Average 1957	50.0	50.0	112.3	105.7	103.3	84.5
			Real Rates (1965 wŏn per dollar)			
Dec. 15, 1953	152.5	152.5	–	–	328.0	248.3
Nov. 10, 1954	104.0	104.0	449.1	427.7	379.2	306.4
Dec. 13, 1954	101.0	101.0	454.4	438.2	299.4	323.6
Jan. 10, 1955	76.3	148.3	391.1	353.8	327.1	266.5
April 18, 1955	74.6	145.2	313.7	193.4	310.4	251.0
Aug. 15, 1955	154.8	154.8	294.1	253.9	248.3	205.0

TABLE 9 (Continued)

	Official Rate	Counter-part Deposit Rate	Japan Export Dollars	Other Export Dollars	U.S. Greenbacks	Military Payments Certificates (MPC)
Average 1956	136.6	136.6	292.3	275.4	263.9	221.3
Average 1957	117.6	117.6	264.2	248.7	243.1	198.8

Source: Same as Table 8.

exchange rates.[3] Throughout most of the period, effective exchange rates were above nominal rates because of tariffs, surcharges, and other pricing measures superimposed on the exchange rate. Estimates of effective exchange rates are given for the 1955 to 1960 period in Table 10. There are no estimates of effective rates available for the earlier years.

The nominal exchange rates in effect on December 15, 1953, are reproduced in Table 9 to provide a basis for comparison. A major devaluation at that time had tripled the official exchange rate, although in real terms it was well below its pre-1950 level. As can be seen, the nominal exchange rate was constant throughout 1954, both for official transactions and for redemption of wŏn advances. As a consequence, the real exchange rate for transactions carried out at the official rate and the counterpart-deposit rate fell sharply, reaching about half the December 1953 level by January 1955.

To counteract some of the effects of severe overvaluation (which, as indicated above, was done as a deliberate policy to increase the real value of the dollars received in exchange for wŏn advances), special export rates were introduced during 1954. One of these was for "Japan Export Dollars." This rate was the dollar rate at which exporters of commodities to Japan could sell their yen to importers. Another "export dollar rate" applied to dollars earned in other markets. The Japan export dollar rate was generally above the other export dollar rate as total imports from Japan were constrained not to exceed earnings from exports to Japan. Both export dollar rates were well above the official rate and the counterpart-deposit rate. Indeed, they were also above the U.S. greenback rate and the military payment certificate rate.

Effective Rates
Although there was no official devaluation between 1955 and 1960, the continuous imposition and alteration of various charges on purchases of foreign exchange and incentives for exports meant that effective exchange rates were altered

frequently throughout the period. For the 1955–1960 period, data are available to provide estimates of effective exchange rates. Table 10 gives estimates of nominal, effective, real (PLD) and purchasing-power-parity (PPP) effective exchange rates.[4]

Charges on imports are examined separately, so that attention can focus on rates applicable to exports at this juncture.[5] As comparison of the export EERs with the export dollar rates in Table 9 indicates, the subsidies to exports were relatively small— probably less than 1 percent of exporters' receipts per dollar of exports. It was these special rates themselves that provided the incentive, such as it was, to export. Manipulation of the Japan dollar and other export dollar meant that export incentives did not depreciate with inflation. The real export EER, given in Table 10, increased throughout the period; it was 223 in 1955, 252 in 1957, and 319 in 1960. Even if the effects of inflation are allowed for, therefore, incentives for export appear to have increased. To be sure, the volume of exports was extremely small even as late as 1960. This resulted from the bias of the regime against exports as reflected in the exchange rate and in the implicit premiums on domestic production of import substitutes.[6]

Charges on Imports
Increasing overvaluation of the exchange rate also occurred with respect to imports. Had there been no quantitative restrictions on international transactions, the inflow of imports would have been significantly greater than even that level which was financed by aid. Exchange control continued, however, so that some commodities were eligible for importation at rates close to those shown in Tables 9 and 10, while other commodities could not be legally imported at all.

Perhaps the most startling feature of Table 10 is the extent to which the EER for imports was below that for exports. This reflects the fact that it was predominantly quantitative restrictions, and not the exchange rate itself, that determined import flows and the level of protection to domestic industries.

TABLE 10 Nominal, Effective, and Purchasing-Power-Parity
Exchange Rates for Imports and Exports, 1955–1960

	1955	1956	1957	1958	1959	1960
A. Official Exchange Rate (wŏn per dollar)	36.8	50.0	50.0	50.0	50.0	62.5
B. Average Export Dollar Premiums (wŏn per dollar)	n.a.	n.a.	n.a.	64.0	84.7	83.9
C. Export Subsidies per Dollar Export (wŏn per dollar)	n.a.	n.a.	n.a.	1.2	1.3	1.2
D. Export Effective Exchange Rate (EER) (A + B + C)	72.2	102.8	108.9	115.2	136.0	147.6
E. Price-Level-Deflated (PLD) EER for Exports (D divided by 1965 price index)	n.a.	n.a.	n.a.	288.7	333.3	326.5
F. Purchasing-Power-Parity (PPP) EER for Exports (E times average price level of trading partners)	223.8	268.3	252.6	280.6	325.6	319.6
G. Tariff Equivalents (wŏn per dollar)	n.a.	n.a.	n.a.	14.4	32.8	37.7
H. EER for Imports (A + C)	42.1	57.3	58.4	64.4	82.8	100.2
I. PLD EER for Imports	n.a.	n.a.	n.a.	160.4	202.9	221.6

TABLE 10 (continued)

	1955	1956	1957	1958	1959	1960
J. PPR EER for Imports	130.5	149.6	135.5	155.9	198.2	216.9

Sources: Kwang, Suk Kim, "Outward-Looking Industrialization Strategy: The Case of Korea," p. 24 in Wontack Hong and Anne O. Krueger, eds., *Trade and Development in Korea,* (Seoul, 1975). Frank, Kim, and Westphal, pp. 70–73, for 1958 to 1960.

Note: See Appendix A for definitions of the exchange rate measures used here.

Charges on imports were significant and the EER was considerably above the nominal exchange rate, increasingly so as the period progressed. Aid was the primary source of foreign exchange, and sales of wŏn for foreign troops was second. It thus fell to the government to allocate foreign exchange for imports both because of exchange control and because the foreign exchange was initially received by the government.[7]

Tariffs and other import charges probably did not affect the composition of imports in any significant way; that was achieved primarily through quantitative restrictions. What tariffs and other charges did do was to absorb some of the differential between import and domestic prices at the going exchange rate.

It will be recalled that tariffs had been imposed in 1949, with rates ranging from 10 to 100 percent. These rates remained in effect, virtually unaltered, until 1957. At that time, they were amended to take into account the altered ability of the domestic economy to provide various goods. In particular, a number of commodities were shifted from the schedules that applied to non-domestically produced goods to the schedules that applied to domestically produced goods. The net result was an increase in the simple average of rates by 4.1 percentage points.

It is difficult to estimate how much tariffs contributed to effective exchange rates. Data on duties collected are available, but the Korean government was shifting fiscal years in 1955 and 1956 (to conform to the American dates) and, as a consequence, tariff collections are reported for a fifteen-month 1955 and an eighteen-month 1957, with no data recorded for 1956. Customs duties and the wŏn value of imports (both in millions of wŏn) reported are listed in Table 11.[8] The large fluctuations in these figures for 1954 to 1957 make one suspect that the change in fiscal year dominates the data. Of course, to the extent that imports of "luxury" goods were increasingly discouraged, a lower average collection rate might have been expected. After 1957, tariff collections as a percentage of imports rose, and tariffs probably constituted an increment to the nominal exchange rate of about 25 percent. Given the vast disparity

TABLE 11 Customs Duties Collected and
Wŏn Value of Imports, 1953–1960

	Customs Duties Receipts	Total Imports	Duties as a Percent of Imports
1953	351	6,200	5.7
1954	999	4,400	22.7
1955	1797[a]	17,000	10.5
1957	1571[a]	22,100	7.1
1958	2969	18,900	15.7
1959	3560	15,200	23.4
1960	5150	21,900	23.5

Note: [a]1955 customs duty receipts were multiplied by 0.8, those for 1957 by 0.66 to make them comparable with import data.

between the official rate and the rate actually paid for imports, it is probable that factors other than tariffs contributed more to the wedge between the EER and the NER than did the tariff, although the relative importance of tariff and other charges undoubtedly varied by commodities. This is seen in Table 10, where the import EER exceeds the NER by 9, 66, and 60 percent in 1958, 1959, and 1960, respectively.

Several other regulations raised importers' costs. In the immediate post-war period, a major means of financing imports was through lending funds to commercial traders. One loan fund was allocated to exporters and raw material users; the other was allocated among industries for imports of capital goods. These loans were supposed to be repaid in dollars within two or three months after receipt. They financed about 75 percent of private imports during the period when the loan fund was in effect—from late 1952 until mid-1954.[9] Initially, when a would-be importer received a loan, he was required to place a wŏn deposit with the Bank of Korea; when the shipping documents were delivered, the borrowers were required to make an additional deposit equal to something between 9 and 23 wŏn per dollar (compared to the official rate of 6 wŏn). After 1953, however, deposits of 20 and 18 wŏn per dollar were required against the first and second loan funds, respectively; applicants for the first fund (for raw materials, and so on) were also required to place the wŏn equivalent of their loan in a one-year time deposit at 4 percent interest.

The loan system was replaced in mid-1954 by an auction system in which blocks of funds were auctioned off to the highest bidder. During 1954, bids ranged from 46 to 69 wŏn per dollar.[10] When it is recalled that the official exchange rate was 18 wŏn per dollar, while export rates were two to three times that level, it is evident that the official exchange rate held little meaning for most categories of transactions other than wŏn advances.[11]

The auction system, too, was abandoned in mid-1955, and foreign exchange for imports was allocated at the official exchange rate (after the devaluation) by lottery among applicants for import licenses. It was during this period, prior to the devaluation, that the premium on import licenses probably reached its peak. While premiums remained substantial over the next several years, it is likely that they were declining as a percentage of import value.[12]

By mid-1957, the lottery system also had been amended; foreign exchange was allocated among bidders on the basis of the amount of national bonds they were willing to buy. Given the vast disparity between the official exchange rate and the value of a dollar of foreign exchange, the implicit cost of national bonds purchased per dollar must have been substantial. Data, however, are not available with which it can be estimated.

The System as of 1960

Over the 1953–1960 period, the official rate was sufficiently unrealistic that it could not reasonably be expected to have allocated foreign exchange. The consequence was an uneasy compromise between allocating foreign exchange at something close to the official rate, letting the fortunate recipients capture the implicit value of the premium, and auctioning off the foreign exchange, with the result that the government earned the implicit premiums. Perhaps of equal importance from the view-point of the effect on domestic producers, the system was in a constant state of flux: no method of allocating foreign exchange remained in effect for more than a year and a half. That there was uncertainty as to the future availability of foreign exchange under any system, auction or lottery, must have increased the protection afforded to the domestic market by the foreign trade regime.

A snapshot picture of exchange rates as of 1960 is provided by the structure reproduced in Table 12. Even that table does not include the tariffs applicable to individual import trans-actions or the subsidies for exports, although the latter were relatively small, as seen in Table 10. As this table indicates, sales of wŏn to U.N. forces continued to be carried out at the official exchange rate, while most other export transactions were effected at a rate twice the official rate. On the import side, the rate structure was more complex. Imports made by the Korean government and some aid-financed imports were effected at the official exchange rate, although in many cases the resale of the commodities was at a price much higher than that paid for the imports, so that the government's profits were reflected else-where. At 80 wŏn per dollar—not significantly above the official rate—cotton and wheat imports were permitted. For most transactions, however, the import rate (before tariffs) was between 110 wŏn per dollar and 134 wŏn per dollar. Thus, the variation in effective exchange rates across commodity categories on the import side appears to have been sizable. While some commodities were imported at rates well below those applicable to export earnings, many others were imported at exchange rates equal to those applicable to exports and, in addition, duties had to be paid on them. Inspection of the range of exchange rates, and recognition that tariffs were high on goods produced domestically, indicate that some imports faced much higher EERs than others. When, in addition, it is recalled that many commodities were either entirely ineligible for importation or subject to severe quantitative restrictions, the apparent anomaly of an import-substitution-oriented regime with a higher EER for exports than for imports disappears.

QUANTITATIVE RESTRICTIONS

As shown above, a number of implicit charges, such as wŏn deposits against foreign exchange loans, applied to particular

TABLE 12 Structure of Nominal Exchange Rates as of
February 23, 1960
(wŏn per U.S. dollar)

Buying	*Selling*
65 (official rate) Applies to voluntary sales of foreign exchange to the Bank of Korea, U.S. off-shore procurement, sales of wŏn to the U.N. forces, and other invisibles.	65 (official rate) Government imports. Aid-financed imports for specified projects.
	80 (official rate plus 15 wŏn tax) Specified aid-financed imports (cotton and wheat) by certain importers (non-project non-end users).
	123.3 (official rate plus 15 wŏn tax plus 43.3 wŏn variable tax) Other aid-financed imports.
133 (transfer rate) Exchange credited to import accounts and sold to importers.	110.5 (official rate plus 15 wŏn tax plus 30.5 wŏn variable tax) Imports financed with exchange supplied by the government.
	134.0 (transfer rate) Imports paid with exchange purchased from holders of Import Accounts.

Source: IMF, *Annual Report on Exchange Restrictions, 1960*, p. 230.

Note: Other rates applied to trade with Japan and missionary transactions. Export subsidies, tariffs, and other charges are not included.

categories of transactions. Despite the multiplicity and complexity of exchange rates, however, they were not the chief instrument employed to contain the demand for imports; quantitative restrictions were dominant.

As previously mentioned, beginning in 1949, quarterly import programs had been published, listing commodities that would be eligible for importation and the value of imports of each category that would be permitted. There were three levels of categories on the lists—section, group, and item—and substitution within groups was possible with permission of the Ministry of Commerce and Industry. This sytem continued until the middle of 1955, although the programs became semi-annual after 1953 and, of course, the list of commodities was altered to suit current conditions.

Any quantitative restriction system under which items not listed are ineligible for importation—a positive list system—is inherently restrictionist, since the import prohibition implicit in failure to list an item provides a high degree of protection. In addition, the facts that would-be importers must apply to a ministry for a license and that the application is reviewed for conformity with the import program provide additional disincentives to importation, both because of the delays and red tape surrounding licenses and because of uncertainty as to whether the license will be issued (either because the commodities applied for might be construed to be outside the list of permitted imports, or because the quota might be oversubscribed and the application either only partially fulfilled or denied).

There is little information about the detailed working of the system prior to the 1955 devaluation. Imports were then partitioned into two groups of eligible commodities—special imports and ordinary imports. The former were eligible for importation only with proceeds from export sales; the latter could be imported with foreign exchange either sold by the government or earned from exporting. Once an item appeared on a list, no license was required. There were a few commodities for which prior approval was still necessary, and implicit prohibition continued to prevail for commodities not listed, although the Ministry of Commerce and Industry was empowered to grant licenses for commodities not listed upon application and approval.[13]

Thus, after 1955, a trader wishing to import commodities listed in the import program first had to apply for foreign exchange. The allocation of this foreign exchange was by lottery. Once the trader had received his allocation, he could import any item on the list of ordinary imports without further paperwork except, of course, for the opening of a letter of credit. The fact that some commodities could be imported only with export earnings heightened the effective exchange rate applicable to those exports; there is no record, however, of the differential value of such eligibility.

The 1955 alterations in the quantitative control system undoubtedly resulted in some liberalization, particularly for those imports no longer requiring licenses. The months prior to August 1955 must, therefore, be regarded as the period during which the constraints of quantitative restrictions reached their height in Korea. After that date, the devaluation, and later the auctions in accordance with bids to purchase national bonds, absorbed some of the excess demand for foreign exchange, thereby doing some of the work in containing demand that had earlier been done by QRs (Quantitative Restrictions).

The 1958–1960 years saw little change in the import regime. As of 1960, imports were divided into two groups of eligible items. The first list was AA (Automatic Approval) and consisted of "essential" commodities. The second list was composed of items deemed "less essential" for which an import license had to be obtained.[14] That the implicit protection accorded to some commodities under the system was very large is evidenced by the fact that, when the 1961 devaluation raised the nominal exchange rate to 130 wǒn and the regime was liberalized, a variable exchange tax was imposed upon imports of those commodities still subject to quota where premiums continued to accrue to import licenses. These rates were in addition to regular tariffs. They ranged from 10 percent to 100 percent. For commodities in the latter category, this must have implied a predevaluation implicit price of 260 wŏn per dollar—well above any of the premium-exclusive EERs given in Table 12.

EXPORTS, 1953 TO 1960

The most important fact about exports in the 1953–1960 period is that they were a relatively unimportant source of foreign exchange. The second most important fact is that they were stagnant, so that there was no apparent basis on which to hope that they might become important. The first assertion is readily documented by the evidence that net capital imports were more than 6 times exports in 1953 and 1954, and more than 10 times exports in every year from 1955 to 1959, reaching a peak of 19 times exports in 1957. That they were stagnant is indicated by the fact that exports were $40 million in 1953 and did not reattain that level until 1961.

It should also be noted that there was no other significant source of foreign exchange that was not associated with aid; private capital flows were negligible, and the services balance was positive only by virtue of the local currency expenditures of the United Nations forces.[15] These two phenomena together underline the central fact about the relationship between aid and trade in the 1950s: not only was aid, and related expenditure by the military forces, the only major source of foreign exchange, but there was also every expectation that it would continue to be so.

These topics will be dealt with later. Here, focus is upon the behavior of exports, such as they were. In terms of the history of modernization of Korea, the topic would be unimportant were it not for later events. A question of considerable importance is why there was such a pronounced change between the insignificance of exports in the 1950s and their role in the 1960s.

Table 13 gives data on the commodity composition of exports for the 1953–1960 period. Several features are noteworthy. First, of course, is the small absolute size of the export sector and its failure to grow. The decline in the value of mineral exports (mostly tungsten) partly reflected a drop in metals prices on world markets, but the volume of exports also behaved erratically. The United Nations provides data on the volume of Korean exports of non-ferrous ores and concentrates, almost all of which must have been tungsten. Exports were 26.9 thousand metric tons in 1953, 27.7 thousand metric tons in 1954, then 37.5 and 38.5 thousand metric tons in 1955 and 1956, followed by 14.5 thousand metric tons in 1957.[16] Second, and related to the first, is the fact that the failure of exports to grow was not the result of a single category's poor performance: there was *no* sector, at least at the level of aggregation shown here, which grew systematically.[17] Had there been consistent growth of some sub-sector, it must have remained very small not to be separated out for special treatment.[18]

Third, Korea was relatively typical for a country with low per capita income in that more than three-quarters of her exports originated in primary products. The percentage distribution of exports is given in Table 14. Minerals were, to be sure, more important than agriculture, but it was the extractive industries, and not manufactures, that provided those export earnings Korea was able to realize. Of course, while Korea was typical of developing countries in having most of her exports originate in primary sectors, she was atypical in that exports constituted such a small fraction of GNP—less than 2 percent for each of the years under review. For most countries, of course, exports are a high fraction of GNP because export earnings provide the foreign exchange that makes imports possible in a relatively specialized economy oriented toward the production of primary commodities. In the Korean case, aid, and not exports, provided the foreign exchange needed to offset the specialization of the economy. Thus, while the small fraction of exports was atypical of poor countries, the dependence upon

TABLE 13 Sectoral Composition of Exports, 1953–1960
($1,000s)

Sector	1953	1954	1955	1956	1957	1958	1959	1960
Rice, Barley, Wheat	–	16	247	–	–	–	775	3,762
Other Agriculture	3,465	2,997	3,462	5,034	2,945	2,868	3,598	3,080
Forestry & Fisheries	1,198	632	416	718	2,294	707	931	2,636
Minerals	29,252	15,009	9,061	14,938	11,506	7,275	8,464	11,372
Processed Foods & Beverages	80	259	405	214	183	1,936	2,383	4,146
Textile Fabrics & Fiber Spinning	2,589	2,693	2,238	2,772	3,260	930	2,165	3,800
Lumber & Plywood	–	–	–	–	41	212	–	–
Chemicals & Rubber Products	–	1,686	247	18	47	10	115	591
Petroleum & Coal Products	1,133	688	387	–	–	297	–	–
Glass & Stone	1,021	17	91	135	–	124	121	–
Steel & Metal Products	–	252	961	1,298	1,195	1,253	479	1,504
TOTAL	39,586	24,243	17,604	25,154	21,521	16,451	19,165	31,833

Source: Wontack Hong, *Factor Supply*, Tables A.11 and A.12.

Note: Miscellaneous Manufacturing, Other Sources, and Unclassifiables are not listed, but are included in the totals.

TABLE 14 Percentage Distribution of Exports 1953–1960

Sector	1953	1954	1955	1956	1957	1958	1959	1960
Agriculture	11.8	15.0	23.4	22.9	24.3	17.4	22.8	21.4
Minerals	73.9	61.9	51.4	59.3	53.5	44.2	44.1	35.7
Total Primary[a]	85.7	76.9	74.8	82.2	77.8	65.9	71.8	65.4
Manufactures	14.3	23.0	25.0	17.7	22.1	34.1	28.2	34.6

Source: Data from Table 13.

Note: [a]Primary includes Forestry and Fisheries.

imports—which showed up in the import/GNP ratio—was not.

THE EXTENT OF IMPORT SUBSTITUTION

The only aggregate estimate of the contribution of exports and import substitution to growth during the 1950s is that of Frank, Kim, and Westphal. They found that the total direct and indirect contributions of export expansion and import substitution in the late 1950s were approximately equal. As would be expected, however, the emphasis on import substitution shows up when the growth of the manufacturing sector is examined separately. For that sector, the direct contribution of export expansion to growth was 5.1 percent, while import substitution accounted for 24.2 percent.[19] These estimates contrast sharply with estimates for the 1960s.

Another comparison is perhaps also instructive. Frank, Kim, and Westphal used Chenery's estimates of the "normal" structure of countries at differing stages of development and capital inflows to contrast Korea's structure with the "norm." While such comparisons are always subject to qualifications, what is perhaps significant is that Korea's share of exports was well below that which would have been forecast, based on 1955 per capita income and the actual capital inflow, regardless of whether Korea is regarded as a large developing country or a large manufacturing developing country. Korea's actual structure in 1955 and 1960, and the "norm" for 1955, are shown in Table 15.[20] As can be seen, exports, manufactured exports, and imports all had shares of GDP well below what would have been forecast on the basis of the structural norms. While this evidence is by no means conclusive, it strongly suggests that the inward-oriented nature of the Korean economy during the 1950s was far greater than would have been expected for Korea's level of development. In the absence of capital inflows, the deviation of exports from their norm would have been even greater than the

TABLE 15   Korea's Actual Structure for 1955 and 1960
versus the "Norm" for 1955

	1955 Actual	1960 Actual	1955 Norm
Per capita GNP	$79	$86	$79
Capital inflow as			
% of GNP	7.7	8.5	7.7
% of exports in GDP	1.7	3.4	9.8
Industry % of GDP	13.0	15.6	17.0
% of manufacturing			
exports in GDP	0.4	1.2	1.4
Imports as % of GDP	10.0	12.7	17.6

data given above suggest: Chenery's equations show that a
country with a per capita income of $79 would be expected to
have imports of about 16.1 percent of GDP.

Despite the emphasis on import substitution in manufacturing
during the 1950s, it is of interest that Korea's share of industry
in GDP fell below that predicted for large manufacturing
countries. While the 1960s were to show a different pattern, one
would hardly have classified Korea as a "manufacturing"
country in the late 1950s. Import substitution clearly failed in
its purpose in the sense that manufacturing had not become a
dominant sector.

Another measure of the extent of import substitution is
provided by the data in Table 16, which gives the ratio of
imports to domestic consumption by sectors for the years 1953
to 1960. Primary Production, Beverages and Tobacco, Leather
and Leather Products, and Printing and Publishing are omitted
from the table, since domestic consumption was supplied by
domestic production in those sectors throughout the period.
Inspection of the data in Table 16 suggests that, regardless of
how much the regime was oriented toward import substitution,
the actual extent of import replacement was relatively small. To
be sure, there are some sectors (such as chemical fertilizers) in

TABLE 16 Sectoral Ratios of Imports to Domestic Consumption, 1953–1960

	1953	1954	1955	1956	1957	1958	1959	1960
Processed Foods	.019	.013	.012	.003	.022	.078	.058	.044
Textiles	.108	.135	.146	.051	.037	.053	.055	.071
Apparel	.015	.003	.004	.018	.013	.044	.008	.004
Lumber, Wood & Furniture	.000	.024	.010	.001	.001	.217	.110	.013
Paper & Products	.428	.372	.414	.181	.208	.475	.341	.304
Basic Chemicals	.145	.127	.162	.092	.077	.414	.310	.303
Coal & Petroleum Products	.152	.066	.089	.010	.045	.600	.561	.336
Glass & Clay Products	—	.066	.047	.065	.003	.140	.116	.024
Fertilizers	1.000	1.000	1.000	1.000	1.000	1.000	1.000	.967
Non-metalic Mineral Products	.083	.030	.028	.009	.016	.179	.054	.073
Basic Metal Products	n.a.	.401	.311	.169	.085	.448	.273	.211
Fabricated Metal Products	.001	.061	.047	.042	.021	.099	.043	.095
Machinery	.388	.359	.236	.222	.007	.405	.441	.450
Electrical Machinery	.155	.413	.291	.122	.194	.549	.388	.435
Transport Equipment	.334	.236	.023	.043	.010	.195	.175	.036
Miscellaneous Manufacturing	.071	.120	.119	.147	.159	.301	.173	.081

Source: Suk Tai Suh, "Import Substitution and Economic Development in Korea," (Mimeo, December 1975), Table 5-2.

which the investment of the 1950s did not bear fruit until the early 1960s, and that is not reflected in the data.

Even taking those factors into account, the data show surprisingly little change in import shares. Such import substitution as did occur is reflected in only relatively small declines in the import-consumption rates, largely because much of the incremental production absorbed excess demand that was initially unsatisfied. That is, foreign exchange availability in the early post-war years was so limited that desired imports exceeded actual imports by sizable amounts. When domestic production increased, it served mostly to meet some of the previously unsatisfied excess demand, rather than to replace imports.

This pattern is reflected in a number of sectors. For example, in Textiles and Apparel there was a considerable increase in production during the 1950s. Yet, in the early post-war years, imports were limited in overall amount. As can be seen, the ratio of textile imports to domestic demand did decline between the 1953-1955 period and later years; however, imports had only accounted for 10-15 percent of domestic consumption prior to 1955, so that the decline could not be very great.[21]

It is generally thought that import substitution was concentrated on light industry, especially consumer goods, and that opportunities for further moves in this direction had been exhausted by the late 1950s. According to Kwang Suk Kim,

> The Republic of Korea ... started out with an industrialization strategy based on a policy of import substitution. She completed the easy import substitution in nondurable consumer goods and their inputs by around 1960. Instead of emphasizing further import substitution in machinery, durable consumer goods and their intermediate inputs, however, Korea changed its industrialization strategy from import substitution to export promotion in the early 1960s.[22]

Suk Tai Suh has provided estimates of the dollar value of production in different sectors over the 1953 to 1960 period.[23] According to his estimates, domestic output in millions of U.S. dollars rose as shown in Table 17. As can be seen, manufacturing

TABLE 17   The Dollar Value of Production
in Different Sectors, 1953–1960
($ millions)

	Primary Industry	Light Manufacturing	Heavy and Chemical Industries
1953	997	417	109
1954	942	564	146
1955	1,106	632	144
1956	1,210	684	151
1957	1,310	732	179
1958	1,318	811	205
1959	1,232	893	264
1960	1,343	976	330

increased in importance relative to primary industries with the value of manufacturing output rising from just over 50 percent of the value of primary production in 1953 to an almost equal amount in 1960.[24]

Suk Tai Suh's data indicate that Heavy and Chemical Industries output increased approximately threefold between 1953 and 1960, while Light Manufacturing output rose two and a half times. A certain amount of the chemical and heavy industry expansion was in consumer goods—coal briquets, toiletries, and so on. Moreover, the absolute increase in output of light industry was greater than that for heavy industry.

Thus it seems reasonable to conclude that it was import substitution, and not export promotion, that was the major source of industrial growth in the 1950s, and that resources were pulled primarily into activities producing for the domestic market.

IMPORTS AND AID

The contribution of aid to Korean modernization had two distinct aspects: 1) aid financed the major part of commodity imports; and 2) negotiations between the aid donors and the Korean government influenced economic policy in a variety of ways.

AID FINANCING OF IMPORTS

Table 18 summarizes aid received by the Korean government from the major donors over the 1953–1960 period and, for purposes of comparison, includes the dollar value of total imports, imports as a percent of GNP in current and constant prices, and the current account deficit as a percent of GNP. There are a number of reasons why the figures must be taken as only approximately indicating orders of magnitude. First, aid did not always finance imports dollar for dollar. Foreign exchange reserves also rose and fell and other components of the current account balance fluctuated. The year 1957 illustrates this, as recorded aid is 87 percent of imports. Despite this, reserves rose $17 million in that year, there was a negative balance on services account, and exports represented more than 5 percent of imports. Second, the overvaluation of the wŏn resulted in an underestimate of the contribution of aid to GNP, and also imports as a fraction of GNP, since imports were valued at their landed cost, and not at the premium-inclusive domestic prices. This can be seen by comparing imports as a percent of GNP in current prices and in constant prices. For years until 1958, measurement at constant (1970) prices shows the share of imports to be greater than does measurement based on current prices. Third, it can be argued that the trade deficit seems a better measure of the importance of aid, since the deficit was possible only because of aid flows.[25] As with the measures given here, however, other factors entered into year-to-year fluctuations in that figure, and it also tends to misstate the economic contribution of aid. The appropriate conclusion is that there is no ideal measure of the importance of aid, but the orders of magnitude represented in Table 18 provide a rough idea of its role.[26]

Regardless of the necessary qualifications, it is apparent that

TABLE 18 Aid Received and its Importance, 1953–1960
($ millions)

	1953	1954	1955	1956	1957	1958	1959	1960
U.S. Bilateral	12.8	108.4	205.8	271.0	368.8	313.6	219.7	245.2
CRIK	158.8	50.2	8.7	0.3	–	–	–	–
UNKRA	29.6	21.3	22.2	22.4	14.1	7.7	2.5	0.2
Total	201.2	179.9	236.7	293.7	382.9	321.3	222.2	245.4
Total Imports	345.4	243.3	341.4	386.1	442.1	378.2	303.8	343.5
Aid as a % of Imports	58.3	73.9	69.3	76.1	86.6	84.9	73.1	71.4
Imports as a % of GNP (current prices)	12.9	7.3	9.8	13.1	12.0	10.7	10.1	12.6
Imports as a % of GNP (constant prices)	n.a.	8.8	11.2	13.0	14.3	11.7	9.3	10.4
Current Account Deficit as a % of GNP	n.a.	6.2	8.7	11.7	10.5	8.7	7.5	9.3

Sources: BOK, *Economic Statistics Yearbook, 1960* and *1974*, and IMF, *International Financial Statistics*, (May 1976).

Note: Imports as a fraction of GNP is calculated including imports of both goods and services; current account deficit as fraction of GNP is imports of goods and services less exports of goods and services at current market prices.

aid financed the vast majority of imports. When it is recalled that dollar receipts by the Bank of Korea against wŏn advances are not included in the aid total, it is evident that intergovernmental relations were a far more important determinant of foreign exchange availability than was policy to promote exports.

As Table 18 indicates, even when aid is valued at the official exchange rate, it accounted for a very large fraction of GNP. If one takes aid as a percentage of imports, and multiplies that by imports as a percent of GNP (to obtain an implied estimate of aid as a fraction of GNP—a not entirely trustworthy procedure), the result indicates that aid was equal to 6–7 percent of GNP in 1953–1954, and rose in relative (and absolute) importance in the 1955–1957 period, reaching almost 14 percent of GNP in 1957.

The macroeconomic implications of this flow are profound: an import surplus of the size financed by aid was strongly deflationary and permitted budget deficits with much less inflationary pressure than would otherwise have resulted.[27] Simultaneously, the additional flow constituted the economy's entire source of capital formation in the early post-war years. Equally important, however, was the nature of the imports provided.

A breakdown of aid financial expenditures on a calendar year basis is given in Table 19.[28] As can be seen, the totals do not quite coincide with those in Table 18, although UNKRA expenditures are included in both. The reason appears to lie both in the discrepancy between authorizations and expenditures, and in the difference between fiscal and calendar years. As can be seen, general program support was far and away the largest aid category; this (and PL 480) was used to finance imports. Even project support, which was much smaller, consisted largely of commodity support.[29] Thus, while project aid made important contributions to specific sectors, the major impact of aid was via the imports financed under it.

TABLE 19 Categories of Aid Expenditures, 1954–1960
($ millions)

| | Supporting Assistance | | Technical | PL 480 | | |
	Non-Project	Project	Support	Title I Sale	Title II & III	Total
1954	74.3	6.0	—	—	—	80.3
1955	168.3	34.8	0.1	10.0	15.9	229.1
1956	220.8	53.1	1.2	37.5	16.8	329.4
1957	207.2	92.6	2.8	30.4	28.3	361.3
1958	163.0	67.2	3.4	38.6	22.3	294.5
1959	148.2	68.8	3.1	12.5	16.9	249.5
1960	160.0	56.3	3.4	32.6	15.1	268.7[a]

Source: USAID to Korea. Data are from Suk Tai Suh's Appendix, Table II-1.

Note: [a]The 1960 total includes $1.3 million from the Development Loan Fund which is not included in any sub-category.

COMMODITY COMPOSITION OF IMPORTS

While the commodity composition of aid-financed goods is available for each separate assistance program (FOA, MSA, CRIK, UNKRA, etc.),[30] no single breakdown for the total of all assistance is available, and the classifications of goods are not the same. For that reason, and also in light of the fact that aid financed such a high fraction of imports, it seems best to examine the commodity composition of total imports. That breakdown is given in Table 20, which shows that food imports were important throughout the period, although less so than they had been during the Korean War. For reasons that will be discussed later, the "manufactures" component, which consisted largely of finished products (but also included such items as newsprint and cement), was also sizable.

Imports provided a large number of items that would otherwise have been unavailable or extremely scarce in Korea. For example, 100 percent of fertilizer availability (included in chemicals) originated through imports until 1960. In the years 1953–1955, over 10 percent of textile and clothing domestic

TABLE 20 Commodity Composition of Imports, 1953–1960 (% distribution)

	Food and Beverages	Crude Materials and Fuel	Total Primary	Chemicals	Manu- factures	Machinery and Transport Equipment
1953	47.6	6.0	55.0	16.4	22.8	3.4
1954	17.3	5.8	24.7	15.2	41.9	13.6
1955	15.9	22.0	37.9	17.5	18.5	16.8
1956	14.0	24.1	38.1	19.3	20.8	11.1
1957	26.0	23.6	49.6	17.4	14.5	9.6
1958	18.4	28.9	47.3	18.1	18.1	9.7
1959	9.0	33.9	42.9	22.6	14.6	13.7
1960	9.2	27.4	36.6	22.2	15.4	11.7

Sources: Wontack Hong, "Trade, Distortions, and Employment," Statistical Appendix, Table C.42, 1955 to 1960; BOK, *Annual Economic Review, 1955* and *Economic Statistics Yearbook, 1953* and *1954.*

Note: An "unclassifiable" category is omitted, and thus totals do not add to 100.

consumption was met through imports. While figures fluctuated from year to year for other commodities, imports constituted more than 20 percent of domestic consumption in one or more years for the following sectors: Other Minerals (1954); Paper and Paper Products (1953 through 1960); Basic Chemicals (1958 to 1960); Coal and Petroleum Products (1958 to 1960); Basic Metals Products (1954 and 1955 and 1958 to 1960); Machinery (1953 to 1956 and 1958 to 1960); Electrical Machinery (1954 and 1955 and 1958 to 1960); and Transport Equipment (1953 and 1954).[31]

It is difficult to transform data on the commodity distribution of imports by sector of origin into meaningful estimates of the distribution of imports among user categories. Some idea of the destination of imports may be gleaned, however, if one is prepared to be heroic and assign Food and Beverages and Manufactures (Standard International Trade Classifications 6 and 8) to Consumer Goods, Crude Materials and Fuels and Chemicals to Intermediate Goods and Raw Materials, and Machinery and Transport Equipment (SITC 7) to Investment Goods. The composition of imports by category of final demand, in percentages, is illustrated in Table 21.[32]

As can be seen, consumer goods—most of which required no further processing in Korea—predominated among imports early in the period under review. Even intermediate goods, which consisted largely of fuel and fertilizer, were goods that required little or no processing in Korea. Perhaps most startling of all, however, is that, in a specialized developing economy which undoubtedly had very little capacity to produce its own producer goods, investment goods constituted less than 14 percent of all imports in each year from 1953 to 1960, and were less than 10 percent in 1953, 1957, and 1958. This reflects in part the conflicts between the ROK and American governments over economic policy.

THE SCARCITY VALUE OF IMPORTS

All pieces of evidence point to the existence of sizable premiums

TABLE 21 Composition of Imports by Category of Final Demand, 1953–1960

	1953	1954	1955	1956	1957	1958	1959	1960
Consumer Goods	70.4	59.2	34.3	34.8	40.5	36.5	23.6	24.6
Intermediate Goods and Raw Materials	22.4	21.0	39.5	43.4	41.0	47.0	48.5	49.6
Investment Goods	3.4	13.6	16.8	11.1	9.6	9.7	13.7	11.7

on import licenses during the 1953–1960 period, but especially in the years 1954 to 1957. The rapid inflation, the overvaluation of the currency, and the licensing system all suggest that the profit accruing to those who could command foreign exchange and import must have included a sizable scarcity rent.

Efforts to obtain data with which to estimate the size of those rents, however, are fraught with difficulty. Ideally, what would be desirable for such estimates would be comparisons of landed cost of individual import items, normal (competitive) wholesale markups, and data on wholesale prices of comparable commodities. In actual practice, the best estimate of landed cost available is unit value; to attempt to convert that estimate into a wŏn figure would already entail large margins of error, and the fact that most commodity categories are not homogeneous makes price comparisons virtually impossible.[33]

Despite all these very real drawbacks, an effort was made to find some price data from which to infer premiums on imports.[34] The outcome of that effort with respect to cement is representative of the best that can be done. Cement is homogeneous, price data are available on a comparable basis, and cement was an import commodity in the early 1950s, an import substitute in the late 1950s, and an export in the 1960s. Price data were unavailable for years prior to 1955, which is unfortunate, in view of the suspicion that premiums peaked at about that time. From 1955 on, the domestic price per ton was divided by the unit value (in dollars per ton) for one ton of imported cement. The computation yielded an implicit exchange rate for each year, which could then be compared to the official exchange rate. The results were an estimated ratio of the implicit exchange rate to the official exchange rate of 2.09 in 1955 (after devaluation), which ratio then rose gradually to a peak of 2.88 in 1958, falling to 2.44 in 1960, 1.34 in 1961, and declining gradually thereafter (reaching 0.89 in 1969). The same general pattern prevailed for the few other commodities for which the exercise was undertaken.

Two things stand out clearly from the results. First, if the

Westphal-Kim estimate of EERs is correct, premiums on cement imports probably equaled one to one and one-half times the cost, insurance, and freight (c.i.f.) price of the import at the official exchange rate, since the excess of EER over the official exchange rate is rather small. Second, the most striking result is that the official exchange rate was well below the implicit rate throughout the 1950s, and it was not until the devaluation of 1961 that the gap began closing. Any examination of the data strongly suggests that the 1950s (at least for 1955 and after) were homogeneous with respect to the relationship between the implicit (premium-inclusive) exchange rates and the official exchange rate. Since premiums of this size have significant implications for the allocative efficiency of the regime, further consideration is given to the issue in Chapter 5.

COUNTERPART FUNDS

When aid-financed commodities are sold in the domestic market, the receipts from their sale are counterpart funds. They represent the domestic purchasing power "counterpart" to the aid flow, and it would be in error to count, as the aid contribution, *both* the inflow of commodity imports *and* the counterpart funds.

Nonetheless, the ways in which counterpart funds are administered can have important implications for resource allocation. On the one hand, since all economic entities are subject to a financial constraint, allocation of funds to one budget rather than another influences the relative abilities of the recipients to compete for real resources. On the other hand, the purchasing power absorbed by imports in effect "frees," for the same level of total expenditures, domestic resources.[35]

It has already been seen that imports, which were predominantly aid-financed, represented a larger fraction of GNP than did investment in the early reconstruction years. It was largely through the allocation of counterpart funds to financing investment that imports filled this role. During the 1950s, central government expenditures by the Republic of Korea were financed by counterpart funds to a very significant degree. In 1957, for example, counterpart funds constituted 53 percent of government revenues, whereas regular sources—mostly taxes—constituted about 34 percent of government revenues.[36]

For that reason, the ROK budgetary process was integrally linked to decisions concerning the allocation of counterpart funds. The American government, however, naturally participated in decisions as to how counterpart funds should be expended. Even when counterpart funds constitute a much smaller fraction of total government expenditures than was the case in Korea, frictions between donor and recipient are bound to arise. When the fraction of ROK expenditures subject to negotiation with American representatives reached the heights that it did, American participation in Korean decision-making became virtually all-inclusive. That leads directly to the final topic of concern during the 1953–1960 period.

THE AID RELATIONSHIP

During the 1950s the Republic of Korea was heavily dependent upon aid, not only for its growth prospects, but even for its day-to-day functioning. Dependence extended not only to imports (including such supplies as gasoline), but also to the expenditure of counterpart funds. American officials were involved because they held the purse strings over virtually all Korean decisions. With such heavy interdependence, it was natural that difficulties would arise between the ROK and American officials.

Several factors accentuated the difficulties in the dependency relationship in the early years. First, and probably most important, aid to an underdeveloped country that was not a colony was without precedent. While the United States had, through the Marshall Plan, participated in the recovery of Europe, that experience itself raised a variety of diplomatic problems from which lessons had not yet been fully learned. The thorny questions associated with negotiating with the government of a country that was heavily dependent and might, if aid were not forthcoming, again be attacked from the north, were probably unresolvable.

Second, and not independent of the first reason, neither American nor Korean objectives seem to have been well defined. The Americans seem to have been uncertain as to whether the objective of aid was to enable self-supporting growth of the Korean economy after a period of time, or whether, instead, the feasible goal was simply to envisage providing support on a continuing basis for a country that could not hope to achieve an economy able to provide a satisfactory standard of living for its people in the absence of an unrequited flow of goods and services. On the Korean side, the situation appears to have been equally unclear. On the one hand, there seems to have been an element of attempting to maximize the volume of aid received, which in turn implied keeping the economy dependent and preventing it from becoming able to generate capacity to earn its own imports. On the other hand, economic growth was an objective of some considerable importance, although it seems generally to have lost out to the aid-maximizing objective in the early years, perhaps because of the same sort of pessimism about the economy's potential that characterized the second American view.

AMERICAN RETURN TO BILATERALISM

As indicated above, a decision that reconstruction, as well as the military effort, should be made multinational had been made in 1950 when the resolution creating UNKRA was passed. During wartime, however, UNKRA was unable to function effectively, because the military command reported to the United Nations via Washington, while the UNKRA link was directly to U.N. headquarters. Nonetheless, during most of the 1950–1953 period, it was anticipated that UNKRA would be the agency responsible for the post-war reconstruction of Korea. To that end, Congress had passed an appropriation for $166 million, subject to the stipulation that the American contribution to total UNKRA aid could be no more than 66 percent of the total.

Although other countries pledged some support to UNKRA, their pledges fell far below the $85 million that would have enabled the entire American contribution to be spent. This combination of events virtually dictated that the aid effort should be essentially bilateral.

As negotiations to end the Korean War were nearing completion, the question of post-war aid became more pressing. Two missions came to Korea to consider the future prospects. On one hand, Robert Nathan Associates were hired by the UNKRA group. They were engaged for a one-year period to study the Korean economy and to draw up a post-war reconstruction plan. The plan was never implemented, although some of its premises influenced subsequent events. In particular, the Nathan group based their planning on the assumption that South Korea's export prospects and comparative advantage lay in agriculture and minerals. The Nathan Plan, in effect, was a five-year program under which Korea would develop her agricultural and mineral exports in order to pay for manufactured imports.[37]

The second mission was sent by President Eisenhower under the leadership of Henry Tasca. This mission was instructed to advise the administration on Korea's future prospects. It appears to have been given six weeks within which to complete this assignment.[38] The Tasca report was responsible for the recommendation that aid be carried out bilaterally under American auspices, although this took over two years to implement. It also presented a more sanguine picture of Korean prospects than was accepted by most observers, thereby implying that a relatively short-lived aid program might accomplish the task.[39]

A division of labor between the multiple agencies operating in Korea in the immediate post-war period appears to have been worked out whereby UNKRA undertook specific industrial projects, while KCAC (Korean Civilian Assistance Commission) operated under the Foreign Operations Administration in social overhead projects and administered the funds formerly allocated through CRIK. In its role as coordinator, FOA and its successor, the ICA, then assumed responsibility for negotiating with the Korean government as to the terms and amount of aid.

It was the FOA, then, that dealt with Korean officials, with veto power over the use of counterpart funds and the power to decide on the level and composition of aid imports. From the American viewpoint, this implied that the entire range of policies affecting the Korean economy were the proper domain of discussion. The American government had been, and continued to be, highly suspicious of Korean inflation and governmental expenditures relative to domestic sources of revenue. It wanted the Korean government to take measures that would increase domestic saving, thus reducing inflation and also "needed" aid. It was generally believed that, as long as inflation continued, American aid should be confined to goods that would relieve inflationary pressure; in particular, finished consumer goods (and not commodities that would require further processing and hence Korean resources) should comprise the bulk of foreign aid to Korea. Thus, American concern with Korean policy was focused largely on domestic resource availability and inflation and its consequences. The American position seems to have been that additional Korean investment without saving would simply fuel inflation, rather than increase the rate of economic growth. It was this view that explains the high fraction of consumer goods imports in the early post-war years.

KOREAN OBJECTIVES AND AID

Whereas the American objectives focused on increasing domestic saving to substitute for aid and to reduce inflationary pressures, the Rhee Government appears to have had three other objectives: rebuilding the capacity that had been destroyed in the war; maintaining a strong military force; and increasing private consumption levels. As Cole and Lyman describe it:

> These objectives called for high levels of investment and of government and private consumption, and they competed with each other

78

for the available supply of resources. In the short run the major means by which resources could be supplemented was foreign assistance. Therefore, the Korean government sought to mobilize additional resources by maximizing such assistance.[40]

Along with this emphasis on maximizing aid, the Rhee Government continued to reiterate its determination that Korea should once again be unified. Whether this was simply political rhetoric or not, it led to a considerable reluctance to undertake some of the investments that would have been warranted had it been accepted that reunification was not likely in the near future. For example, projected demands for electric power hinged crucially upon the assumptions made with respect to the availability of power from the north. To plan power production on the assumption that there would be no supplies from the north would have belied the government's insistence on the reunification goal.

Out of all this came what Cole and Lyman term an emphasis on short-run objectives:

During the years when Korean leaders were concentrating on augmenting the supply of immediately available resources by obtaining additional aid, they were less concerned with the longer-run effects of the uses to which those resources would be put. They were trying to cope with immediate or short-run problems of security, hunger, and survival, rather than with the future growth of output.[41]

NEGOTIATIONS OVER AID LEVELS

In light of American ambivalence as to aid objectives and the Korean determination to maximize aid levels as a means of reconciling their competing objectives, it was inevitable that bargaining over aid levels should have been accompanied by conflict.

On the Korean side, the objective seems to have been to leave the gap between demand and supply to be filled by aid as large as possible: the exchange-rate overvaluation was perpetuated largely for this reason; inflationary financing of a relatively high level of government expenditures further contributed. Moreover, there was little effort to increase domestic savings, which meant that virtually all gross domestic capital formation was financed by aid; with aid at around 10 percent of GNP, gross domestic capital formation averaged 11.9 percent of available resources (including aid) during the 1953–1957 period.[42]

On the American side, pressure was therefore brought to bear to alter the exchange rate, increase taxes, and reduce deficit financing. Both the 1953 and the 1955 devaluations seem to have been undertaken reluctantly at the urging of the American negotiators as a precondition for receiving aid.

Under these circumstances, bargaining over aid levels would in any event have been difficult. But, as described by Cole and Lyman, the situation was further confounded by the negotiation format which started with agreement upon a target rate of growth:

> Unfortunately, quite often this issue was posed, by both Korean and American officials, in terms of a direct conflict between domestic resource mobilization and foreign assistance. Conceptually, a target rate of growth and "required" military expenditures were assumed, so that any additional resources that could be diverted from domestic consumption were expected to be matched by a reduction in external aid.[43]

The outcome of these negotiations was, as is evident from Table 18, a rising level of aid from 1953 to 1957. The resulting increased flow of imports enabled satisfactory increases in output over the period from 1953 to 1957. It was only in 1957 that the American authorities began making clear that aid levels would decline and that Korean policies would have to alter. It was the policy changes that accompanied the recognition that aid could not grow indefinitely that demarcate the reconstruction period, 1953–1957, from the characteristics of the economy over the 1958–1960 period.

During these later years, the Korean government undertook a stabilization program, which is what accounts for the drastic reduction in the inflation rate in 1958. The chief thrust of the program was a sharp reduction in the rate of increase in the money supply. The extent to which aid bargaining was instrumental in bringing about the stabilization policies is not clear. What is clear is that both the more restrictive monetary policies and reduced import levels in 1958 that resulted from the aid cut had direct repercussions on the level of economic activity, especially on the manufacturing sector. The 1958 to 1960 period was consequently characterized by stagnation of output. This phenomenon, in turn, forced recognition that aid levels would decline in the future, and that stagnation would continue if the policies then being pursued with respect to the trade-and-payments regime persisted.

THREE

The Transition to an Export-Oriented Economy

The period from May 1960 to 1965 constitutes a time of transition during which the entire orientation of trade and exchange-rate policy shifted. The Korean economy was restructured toward export promotion and away from the earlier emphasis on import substitution. The result was the start of exceptionally rapid growth of exports, the beginning of private capital inflows into Korea, and also the continued diminution of both the absolute and the relative importance of aid. For all these reasons, the period is of particular interest both in the context of understanding Korean modernization and also as a case of an exceptionally sharp and successful change in policies. In the latter regard, the Korean policy switch is perhaps the most dramatic and vivid change that has come about in any developing country since World War II. The lessons that emerge from it are therefore important, not only for understanding Korean modernization, but also for other countries.

THE EVOLUTION OF TRADE AND
EXCHANGE-RATE POLICIES

It is useful to begin with a chronology of the policy changes that were effected over the transition period. Then exchange rates, quantitative controls, and export incentives are examined in more detail.

CHRONOLOGY OF POLICY CHANGES

In some respects, it is artificial to pick any particular date as the beginning of the transition. It will be recalled that the relative importance of quantitative restrictions had already peaked in 1955 and that the degree of discrimination against exports was probably at its height at that time. Thereafter, some export incentives were added to the system, although the continuing overvaluation of the exchange rate meant that those measures merely offset some of the disincentives implicit in the exchange-rate regime.[1]

The first major step (after the 1955 reforms and the 1957–1958 stabilization program) in altering the orientation of the regime was undertaken as part of the reforms inaugurated by the Chang Myon civilian government that came to power after the student revolution of May 1960. In January and February 1961, there were two devaluations which were intended to unify the exchange rate and to reduce, if not eliminate, the degree of currency overvaluation.

The military government that came into power in April 1961 continued to pursue policies supportive of exchange-rate unification, increased the scope of export incentives, and liberalized and simplified the remaining quantitative controls over imports. However, it did not continue the restrictive monetary and fiscal policies of the stabilization program and, indeed, adopted fairly expansionary measures. The result was a sharp acceleration in the rate of inflation and, with that, an increase in the demand for imports at the fixed nominal exchange rate. The authorities were obliged, by 1963, to intensify once again the use of quotas and quantitative controls over purchases of foreign exchange.

This represented, in effect, a setback in the move toward liberalization. The disincentive to exports that would otherwise have resulted, however, was largely offset by the introduction of another export-import link system which permitted exporters to use their foreign exchange earnings to import commodities not otherwise legally importable. This move transferred to exporters the premiums implicit on import licenses and served to maintain export incentives and to mitigate the effect of increasing currency overvaluation upon exports.

The final steps in the transition to a consistent export-promotion strategy were taken by the government after the elections early in 1964. Prior to that date, efforts had been made to encourage export promotion, but circumstances had prevented the government from carrying out consistent policies. Thus, inflation had eroded part of the incentives provided, and the payments regime was of varying restrictiveness in response to balance-of-payments pressures. Nineteen sixty-four marks the watershed, after which date export-promotion policies were deeply embedded and consistently administered.

In May 1964, a sizable devaluation was announced. Accompanying it, the import regime was once again liberalized and export incentives were increased. Simultaneously, a number of monetary and fiscal reforms were undertaken which contributed importantly both to the government's ability to maintain the real exchange rate and to the success of the policy initiatives. Thereafter, fluctuations in the real exchange rate for exporters were substantially diminished, and policy changes were fewer and less significant than had earlier been the case.

From May 1960 to 1964, therefore, the transition to an export orientation was marked by a number of twists and turns, as this brief outline of major policy changes indicates. In the more detailed discussion of components of policy that follows, it should be borne in mind, however, that, despite changes in governments, switches in import liberalization, and alterations in the specifics of policy, there does not appear to have been any deviation from an increasing commitment to encourage exports. It was this commitment that appears to have been the underpinning for most of the changes in the trade-and-payments regime which took place during the 1960–1964 period. The details of policies that were adopted appear to have resulted in large part in pragmatic response to the fortunes of exports: when export performance was deemed satisfactory, policies were left unaltered; when, however, it appeared that export growth was faltering, changes were instituted until satisfactory performance was again observed.

Kwang Suk Kim has provided an assessment of the factors that motivated the policy switch toward an export orientation:

> First, the economic growth performance in the late fifties and early sixties was frustrating to both policy-makers and the people, since the possibility of rapid growth through import substitution seemed nearly exhausted by that time. By around 1960, Korea had virtually completed import substitution in nondurable consumer goods and in the intermediate products used in their manufacture. A growth strategy concentrating on import substitution in machinery, consumer durables and their intermediate products did not seem to be an appropriate alternative because of the limitations imposed by the smallness of the domestic market and the large capital requirements. Secondly, Korea's natural resource endowment is so poor that an alternative development strategy based on domestic resource utilization was inconceivable. Thirdly, U.S. assistance, which financed most of the post-Korean war reconstruction, started to gradually decline in the early sixties. Faced with this reduction in foreign aid, Korean policy makers had to seriously consider an alternative source of foreign exchange to meet the balance of payments difficulties. Fourthly, the availability of a well-motivated labor force with a high educational level and relatively low wages provided the country with a comparative advantage in exporting labor-intensive goods. Lastly, one should mention the determination of the leadership to attain a high rate of growth, and a virtual lack of constraints on the ability to make decisions and to carry them out.[2]

As Kim's discussion indicates, the export-promotion policies of the government were adopted as a means, not as an end. The belief that an export-oriented policy would result in significantly better growth performance underlay the switch to exporting. The fact that the growth rate did increase reinforced the commitment to an export-oriented strategy.

EXCHANGE RATES

Table 22 extends the data on exchange rates given in Table 10 to the period 1960–1965. The effects of the two major devaluations can readily be seen in Line A. Whereas the price level rose 13 percent between 1960 and 1961, the official price of foreign exchange increased 104 percent. This large a change absorbed a great deal of the excess demand for imports and the scarcity value of import licenses. As a consequence, the official exchange rate became much more important as a factor in influencing the volume of exports, imports, and other international transactions.[3] Moreover, an increase in the real price of foreign exchange of that magnitude must have absorbed much of the premiums on imports subject to quantitative control while simultaneously increasing the prices of some commodities whose importation had been relatively liberally permitted under the prior regime. Considerable unification of implicit, as well as explicit, exchange rates resulted.

The second devaluation in 1964, from 130 wŏn to 256 wŏn per dollar,[4] was not proportionately as large. Moreover, the intervening inflation had reduced the real price of foreign exchange so that the 1964 devaluation really served to restore the real exchange rate to its 1961 level.

Unlike the 1961 situation, however, fiscal and monetary reforms were undertaken in conjunction with the 1964 devaluation to try to assure future constancy of the real rate. By March 1965, in fact, the wŏn rate was allowed to float in a further effort to maintain its real value and provide assurance to those engaging in international transactions, and especially in exporting, that the new real rate was not simply temporary.

A second difference between the two devaluations was in the relative importance that attached to the official exchange rate prior to each. Thus, while the 1961 devaluation increased the

TABLE 22 Nominal, Effective, and Purchasing-Power-Parity Exchange Rates for Exports and Imports, 1960–1965

	1960	1961	1962	1963	1964	1965
A. Official Exchange Rate (wŏn per dollar)	62.5	127.5	130.0	130.0	214.3	265.4
B. Average Export Dollar Premiums (wŏn per dollar)	83.9	14.6	—	39.8	39.7	—
C. Export Subsidies per Dollar Export (wŏn per dollar)	1.2	8.5	21.5	19.6	27.4	39.2
D. Export EER (A+B+C)	147.6	150.6	151.5	189.4	281.4	304.6
E. PLD EER for Exports (D divided by 1965 price level)	326.5	294.1	270.5	280.6	309.6	304.6
F. PPP PLD EER for Exports (E times average price level of trading partners)	319.6	289.1	264.0	275.8	305.0	304.6
G. Tariff Equivalents (wŏn per dollar)	37.7	19.5	16.4	18.1	32.7	27.7
H. EER for Imports (A+G)	100.2	147.0	146.4	148.1	247.0	293.1
I. PLD EER for Imports	221.6	287.1	261.4	219.4	271.7	293.1
J. PPP EER for Imports	216.9	282.2	255.1	215.7	267.6	293.1

Source: Frank, Kim and Westphal, pp. 70–73. See Appendix A for a list of the symbols used and their definition.

price of foreign exchange from 62.5 to 130 wŏn per dollar, there was a large offset on the export side as other export incentives were equivalent to an additional 85 wŏn per dollar in 1960 and only 23 wŏn per dollar in 1961. Consequently, the wŏn receipts of the average exporter rose only from 147.6 to 150.6 per dollar despite the devaluation. In real terms, there was a decline in the value of a dollar of exports from 1960 to 1961. By contrast, in 1964 export subsidies constituted a smaller fraction of the total incentive to export, so that the real value of a dollar's proceeds in fact increased with the devaluation.

Several other features of exchange rates in the 1961–1964 interval should be noted. First, measures were taken to insulate the real export rate from the effects of inflation. Despite inflation rates of 9 and 20 percent in 1962 and 1963 respectively, the erosion that would otherwise have occurred in the domestic purchasing power of dollar export earnings was largely offset by increased premiums in 1963 and subsidies in 1962. The result was a decline of 8 percent in 1962 in the real value of a dollar's receipts, but an increase of 4 percent in 1963. For exports, therefore, the various incentive schemes provided a buffer against the effects of inflation at a fixed exchange rate. Indeed, the 1964 devaluation was enough to raise the real export rate above its 1961 level, although it remained below its 1960 level.

Second, the situation with regard to imports was the opposite of that for exports: though there was some reduction in the average tariff-equivalent per dollar of imports at the time of devaluation, it was insufficient to offset the major part of the impact on the real cost to the importer of a dollar's worth of goods. The PLD EER for imports therefore rose by 30 percent. Almost all of this increase must have resulted in a reduction in the value of import licenses to their recipients; to the extent that it did not, there was undoubtedly some unification of implicit exchange rates across various import categories.

In further contrast to the treatment of exports between devaluations, the EER for imports hardly changed between 1961 and 1963, so that the real cost of a dollar of imports fell substantially. By 1963, indeed, it was below the level that had prevailed in 1960 prior to the devaluation. Thus, while the exchange-rate change may have resulted in increased reliance upon the pricing mechanism to allocate foreign exchange in the period immediately after devaluation, inflation prevented continuation of that function. Whatever reduction initially took place in the variation in premium-exclusive effective exchange rates, there was a sizable offset by 1963, as the importance of quantitative controls once again increased.

QUANTITATIVE RESTRICTIONS

The 1961 devaluation was intended to reduce, if not virtually eliminate, reliance upon QRs. However, the transition from multiple exchange rates to a unified exchange-rate system[5] meant that there was a variety of commodities for which the gap between domestic price and landed cost would widen unless imports were allowed to increase sharply. Rather than accept that outcome, the government revised the quantitative control system in a manner to be described below and, in addition, levied special tariffs to absorb the differentials between landed cost and domestic price for a wide range of items. About 700 commodities subject to import quota were placed in four separate categories, with special tariff rates (over and above the regular tariffs to which they were subject) of 100, 50, 30, and 10 percent. Commodities were, thereafter, reclassified among these categories as the domestic-foreign price relationship altered. The resulting system was, in consequence, a hybrid, as the import-control regime affected the quantity of imports but did not, in any significant way, generate large windfall gains to recipients of import licenses.

The control regime itself was revised twice during 1961. By the end of the year, there were three categories of commodities: 1) items that could be imported without any prior approval (automatic approval—AA imports) of MCI; 2) commodities that could be imported only after official approval had been obtained; and 3) prohibited items. These categories differed from those in the first half of the year in that the AA classification had earlier been divided into commodities importable with any source of foreign exchange and commodities importable without prior approval only when financed by export earnings.

Within this framework, further liberalization of the import regime could readily have been accomplished by removing commodities from the prohibited and restricted list and shifting them to the AA list. Conversely, the regime could become more restrictionist by a transfer of commodities in the opposite direction. In fact, there was a tendency toward increased liberalization during 1962, but the shortfall in foreign exchange availability occasioned by the declining levels of aid and the increased need for foreign exchange to counteract the poor harvest brought about an abrupt reversal of that trend.

Table 23 gives an indication of the behavior of the trade regime over the 1961–1965 period. The trend toward increasing restrictionism after the second half of 1962 is immediately apparent. By the second half of 1963, the number of AA import items declined to less than 10 percent of its level a year earlier, while the number of commodities requiring approval increased drastically. To be sure, a count of the number of items is not necessarily proof of greater restrictiveness, since it might be possible for fewer commodities to be subject to AA licensing, but the ease with which ministerial permission was granted could have increased for restricted items. There is no evidence, however, that this was the case. And, as will be seen below, the total value of imports fell sharply against the background of domestic inflation, further reinforcing the view that the restrictiveness of the import regime increased markedly after 1962.[6]

The second major devaluation took place in May 1964. Details of the control program for 1964 are not available, but the data on total importable items given in Table 23 strongly suggest that the regime had become fairly restrictive by that time. Indeed, when the devaluation occurred, a new Temporary Special Tariff Law, similar to that which had absorbed the differential between domestic price and landed cost after the 1961

TABLE 23 Number of Items in Each Import Category, 1961–1965

	Automatic Approval	Restricted	Semi-Restricted	Total Importable	Prohibited
First half 1961	1,237 (309[a])	35[b]	—	1,581	305
Second half 1961	1,015	17	—	1,132	355
First half 1962	1,195	119	—	1,314	366
Second half 1962	1,377	121	—	1,498	433
First half 1963	776	713	—	1,489	442
Second half 1963	109	924	—	1,033	414
First half 1964	n.a.	n.a.	n.a.	1,124	617
Second half 1964	n.a.	n.a.	n.a.	496	631
First half 1965	1,447	92	19	1,558	624
Second half 1965	1,495	124	4	1,623	620

Source: Frank, Kim, and Westphal, p. 45. Their data for 1961 and 1963 are based on the original programs, while for the other periods, breakdowns are based on realized figures.

Notes: [a]The total number of AA items for the first half of 1961 was 1,546 of which 309 were eligible only when financed by export earnings and 1,237 were accorded AA treatment regardless of the source of foreign exchange.
[b]For the first half of 1961, restricted items were eligible for importation only with export earnings. Thereafter, restricted items refer to those commodities which could be legally imported only after ministerial permission was obtained.

devaluation, was put into effect. Unlike the earlier law, however, there were about 2,200 commodities affected (compared with 700 in 1961) and there were only two categories with rates of 90 and 70 percent.[7]

After 1964, liberalization of the import controls proceeded rapidly, as can be seen in Table 23. Even in the first half of 1965, the number of AA items exceeded the maximum that had earlier been reached. As foreign-exchange receipts grew rapidly during 1965 and subsequent years, the degree of liberalization that was achieved in 1965 was maintained, and even, on occasion, extended.

Thus, there is a second major contrast between the 1961 and 1964 devaluations. Whereas the liberalization following the 1961 devaluation was fairly short-lived, that following 1964 was far more pronounced and sustained. That was possible largely because the real exchange rate was maintained after 1964, which had not been the case after 1961.

EXPORT INCENTIVES

As already seen, the 1961–1965 period was the time when the commitment to carry out an export-oriented policy was transformed into successful export performance. The means chosen to encourage exports varied pragmatically in accordance with the degree of success then being achieved in exporting. By 1964–1965, the system of export incentives that was to be in effect during the following decade was fairly well established.

Table 24 provides a list of the major export incentives that were used from 1950 onward, along with the dates for which each type of incentive was in effect. Some of the incentives, of course, served merely as an offset to the disincentive for export that the trade regime would otherwise have provided. Tariff exemptions on imports of raw materials, for example, would not by themselves constitute an "export incentive" but would merely serve to enable Korean producers to compete in international markets without negative effective protection.[8]

Before a discussion of the nature of each type of export incentive, a few observations on the entire list are in order. First,

TABLE 24 Types of Export Incentives and Dates of Operation, 1950–1975

Type of Export Promotion Scheme	Dates Applicable
Tariff exemptions on imports of raw materials and spare parts[a]	1959–1975
Tariff and tax exemptions granted to domestic suppliers of exporting firms	1965–1975
Domestic indirect and direct tax exemptions[b]	1961–1972
Accelerated depreciation	1966–1975
Wastage allowance subsidies	1965–1975
Import entitlement linked to exports	1951–1955, 1963–1965
Registration as an importer conditional on export performance[c]	1957–1975
Reduced rates on public utilities	1967–1975
Dollar-denominated deposits held in Bank of Korea by private traders	1950–1961
Monopoly rights granted in new export markets[d]	1967–1971
Korean Trade Promotion Corporation	1964–1975
Direct export subsidies	1955–1956, 1961–1964
Export targets by industry	1962–1975
Credit subsidies	
Export credits	1950–1975
Foreign exchange loans	1950–1954, 1971–1975
Production loans for exporters	1959–1975
Bank of Korea discount of export bills	1950–1975
Import credits for exporters	1964–1975
Capital loans by medium industry bank	1964–1975
Offshore procurement loans	1964–1975
Credits for overseas marketing activities	1965–1975

Source: Frank, Kim, and Westphal, p. 40, covering the period to 1972. Data were updated to 1975 on the basis of information supplied by Korea Development Institute.

Notes: [a]Tariff exemptions were shifted to a rebate system in July 1974.

TABLE 24 (continued)

---

[b]Direct tax exemptions were abolished in early 1973.

[c]The value of exports required to obtain an importer's registration was gradually increased. In 1958, it was $10,000. By 1970, the minimum required export value was $300,000.

[d]Authority was granted in 1962, but was virtually unused until 1967.

some of the incentives listed in Table 24 were included in the computation of effective exchange rates for export. For example, the value of direct export subsidies per dollar of exports is included in the export EERs. The values of some other incentives, such as the wastage allowance subsidies and most of the export-import link schemes, could not be calculated and therefore were not included in estimates of EERs. The result is that the estimates of export EERs given in Table 22 are probably lower-bound indications of the extent to which exports were encouraged by the regime. Moreover, as the long list of types of export incentives and of the dates at which they came into effect suggests, it is probable that the value of export incentives not included in the EER computations increased over time. Second, as inspection of Table 24 shows, a majority of the incentives were already in effect in one form or another by the early 1960s, and no significant new types of incentives were introduced after 1967. The structure of export incentives was therefore stable by 1965, although the relative importance of different incentive schemes varied from time to time and also from commodity to commodity. Third and finally, even a simple listing of the types of schemes employed to encourage exports gives some idea of the extent of the commitment of the government to the export-promotion effort, and also of the pragmatic way in which new schemes were introduced. As those urged to export protested at various disabilities or disadvantages, means were found for removing such disadvantages; when exports lagged, new incentives were introduced or the value of existing incentives increased in order to spur export performance.[9] The incentives provided the authorities with flexible tools with which they could induce the private sector to perform to the extent desired.[10] Perusal of the list is indicative of the extent to which the entire machinery of government became oriented to the attainment of export goals.

Some of the incentives listed in Table 24 are self-explanatory and others have been discussed before. Tariff exemptions on raw materials and spare parts require little comment. Most countries have one scheme or another under which tariffs paid on imported inputs are rebated to the exporter after he has processed and shipped the goods. The Korean system went beyond this by exempting exporters from paying duties in the first place. If, at a later date, exports were less than expected, the importer was expected to pay duties on the difference. In addition to exemptions for direct exporters, the Korean incentive system provided greater inducements than usually encountered by also exempting domestic suppliers of exporters, starting in 1965.

Domestic indirect and direct tax exemptions were introduced as part of the 1961 measures, and were probably of considerable importance. At that time, exports were exempted from the domestic commodity tax and exporters from the business activity tax. In addition, exporters were permitted to reduce their income tax liabilities 30 percent on income from exports and 20 percent on sales to tourists and the U.N. Command. These reduction rates were changed to 50 percent for both categories of transaction in 1962, and remained at that level until their abolition early in 1973. Accelerated depreciation provisions, which did not apply to production for domestic sale, provided yet another tax inducement for exporting.

Wastage allowances are a form of export subsidy that have led more than one observer to question the efficiency of the export drive. Wastage allowances were set as a proportion of required inputs that exporters were allowed to import, over and above established needs, per unit of output. If, for example, it was agreed that producers of a particular commodity required $0.40 of imported intermediate goods per dollar of exports, a wastage allowance of, say, 25 percent might be established, thereby permitting the exporter to import $0.50 of the inputs per dollar of exports. Theoretically, the wastage allowance was designed to cover that fraction of inputs which might be defective, broken in handling, or embedded in commodities whose specifications did not meet quality control. In practice, wastage allowances apparently exceeded any reasonable estimate of genuine wastage. Many of the imported intermediate goods were not otherwise eligible for importation for the domestic market, and they could be legally resold. The result was that the wastage allowance enabled many exporters to earn an additional profit, either by using the excess intermediate goods to produce for the domestic market and sell at a high price, or else by selling the intermediate goods to other producers for a price far in excess of their (duty free) imported price.

The facts that wastage allowances overstated inputs into export production and that exporters had an incentive in any event to overstate their requirements of imported intermediate goods have had consequences for resource allocation. These provisions also created an artificial incentive to employ imported, rather than domestic, intermediate goods and, in so doing, encouraged activities with lower domestic value added than might have been induced under an alternative export-promotion scheme.[11] For this and other reasons, estimates of net exports, which are the appropriate measure of the importance of exports to the economy and which should be used in estimating the growth of exports, are probably biased and subject to a higher margin of error than would otherwise be the case. There is reason to believe that domestic value added in exports is probably understated in the statistics because imported inputs are probably overestimated. For purposes of estimating growth rates, however, there is no basis to determine whether the overestimation increased or decreased over time. This phenomenon influences estimates of employment in exports and a number of other key statistics.

Import entitlements linked to exports, or export-import link schemes, have already been discussed in earlier chapters. These schemes represented a means by which the scarcity value of imports (at an overvalued exchange rate) could be transferred to exporters, thereby maintaining the incentive to export in the face of an unrealistic exchange rate.

An interesting form of export incentive not often found among developing countries but used earlier by Japan arose in the provision, starting in 1957, that only those whose export performance met certain targets could become registered importers. This provision, like the link schemes, tended to transfer the implicit value of foreign exchange to the exporters who were, in fact, earning it. Interestingly enough, one could register as an exporter with a smaller volume of exports than needed for registering as an importer. In 1959, for example, minimum exports required to register as an exporter were $20,000, while minimum exports required to be permitted to be registered as an importer were $100,000. Once a registration certificate was obtained, it did not automatically remain valid. Rather, export targets were raised for each successive year, and they had to be met in order for an importer's registration to remain valid. As with so many other types of provisions, it is difficult to estimate how important this type of provision was as an incentive to exports. It does illustrate the extent to which all were goaded to perform well and to improve upon whatever performance had gone before.

The next two incentives listed in Table 24 require little comment. Reduced public utility rates obviously made exporting relatively more profitable. The dollar-denominated deposits in the Bank of Korea were an inducement only in the 1950s when severe exchange control was in effect. After 1961, with unification of the exchange rate, that incentive was no longer operative. Granting of monopoly rights to the first firm to enter a new market was used for a brief period, from 1967 to 1971, as a further stimulus to exports.

The Korean Trade Promotion Corporation, or KOTRA, was established in 1964, and was designed to assist exporters with marketing activities. It has been an active organization, and has undoubtedly been important in assisting exporters in expanding into new markets and meeting quality and other technical requirements associated with marketing.

Direct export subsidies, as already mentioned, were employed to offset the increased overvaluation of the wŏn that followed after the 1955 and 1961 devaluations. The value of those sub-sidies was included in the computation of EERs for exports, and little further comment is required. Unlike many of the other export incentives, the value of the subsidies can be estimated with fair accuracy. The only point that should perhaps be made is that the export EERs given in Tables 10 and 22 represent an average of the rates applicable to individual items; in fact, subsidies were extended to different products at varying rates and were by no means uniform.

The use of export targets started in 1962 and has played an important part in export-promotion policy since that time. Although a plan was in effect in 1962, the targets set forth in the First Five-Year Plan were well below performance and they therefore had little effect. However, starting with 1962, annual export targets were set, each target exceeding the realized level of the prior year by a sizable amount. It was these annual targets that were operationally significant.

The targets were implemented in a variety of ways. In the first instance, fulfillment was the responsibility of the Ministry of Commerce and Industry (MCI), and an "export situation room" was established to monitor export performance.[12] Tar-gets were assigned to industrial associations, firms, and regions. When exports were at or above their target levels, few changes were initiated. If, however, exports began lagging for a particular sector, efforts were initiated to rectify the situation. Measures extended all the way from threats (and presumably implementa-tion) of sanctions to provisions of additional incentives and government measures to remove bottlenecks.[13] In later years, the political imperative of meeting targets resulted occasionally in such practices as the alteration of dates of exports, the speeding up of shipments, and other devices to assure that the statistics accorded with the target.

The last category of incentives—credit subsidies—reflects both the degree to which the Korean government emphasized its export-promotion goals and also the fragmented state of the Korean capital market. Because there was an excess demand for credit throughout most of the period, exporters were favored not only with lower interest rates but also with preferential access to loans. The fact that loans were extended, even without interest subsidy, at lower rates of interest than borrowers were willing to pay implies that there was an element of subsidy, additional to the lower interest rates, in the preferential status of exporters. That element is not included in the data in Tables 9 and 23. The interest-subsidy values recorded there reflect simply the lower interest charges to exporters. Interest-rate subsidies were, initially, much smaller than other forms of subsidy payments, totaling 255 million wŏn in 1962 compared with 310 million wŏn internal tax exemptions and 566 million wŏn direct subsidy payments. Over the next two years, however, credit subsidization became increasingly important as an incentive for exports. In 1964, interest-rate subsidies constituted more than one-fifth of total export subsidies; they remained at approximately that fraction of the total thereafter.[14]

Some part of the interest rate subsidies probably served to offset imperfections in the Korean capital market and thereby enabled improved resource allocation and the success of the export-promotion strategy. But there can be little doubt that there were also less desirable effects and that the availability of subsidized credit induced the use of more capital-intensive techniques than were probably optimal. The efficiency of resource allocation resulting from the trade regime is considered in Chapter 5.

EXPORT PERFORMANCE

The success of the export-promotion drive was truly phenomenal.

While it is impossible to provide a precise assessment of the degree to which the export incentives were responsible for that performance (as contrasted with the alternative hypothesis that Korean development had proceeded sufficiently so that conditions were in any event right for an export boom), there is no question that export performance exceeded even the most optimistic expectations.

Table 25 gives the sectoral composition of exports for the 1961–1965 period. It will be recalled that exports had stagnated in the 1950s, failing to reattain their 1953 dollar value until 1961 (see Table 13). The first, and most obvious, change in the 1960s was the reversal of the downward trend and the growth of exports. From $31.8 million in 1960, they grew to $38.6 million in 1961, $54.8 million in 1962, $87.0 million in 1963, $118.9 million in 1964, and $175.0 million in 1965. This represented a sixfold increase in export earnings in the unbelievably short space of five years. Moreover, the rate of growth of exports appeared to be accelerating over most of the period: export earnings increased 21 percent in 1961, 42 percent in 1962, 58 percent in 1963, 37 percent in 1964, and 47 percent in 1965.

The most striking feature of the data in Table 25 is the fact that export growth was an across-the-board phenomenon: only Coal Products failed to maintain the export levels of the 1950s. Every other sector contributed significantly, and only two primary-based sectors did not more than double export earnings between 1961 and 1965. Three sectors which already had relatively large exports in 1961 grew at exceptionally rapid rates: Textiles, Lumber and Plywood, and Steel and Metal Products. These three sectors accounted for $83.6 million of the total increase in exports of $136 million. In each of these sectors, moreover, the 1965 level of export earnings exceeded the 1961 level by a factor of more than ten. No other sectors had comparable records except for two whose 1961 and 1965 earnings were negligible (Leather Products, and Glass and Stone). Thus, whereas the three sectors had accounted for 18 percent of total exports in 1961, they accounted for almost 52 percent of total export earnings in 1965.

TABLE 25 Sectoral Composition of Exports, 1960–1965
          ($1,000s)

	1961	1962	1963	1964	1965
Rice, Barley, Wheat	508	8,960	809	2,352	3,242
Other Agriculture	5,078	2,998	4,778	5,071	6,019
Forestry and Fisheries	1,652	4,759	5,812	5,450	8,343
Minerals	14,812	12,252	15,177	20,016	23,764
Processed Food and Beverages	5,962	8,734	12,344	20,499	18,807
Textile Fabrics, Products and Fiber Spinning	4,189	7,623	17,613	32,744	54,553
Lumber and Plywood	1,217	2,289	6,309	11,421	18,177
Wood and Paper Products	114	112	117	372	517
Leather Products	1	2	1	74	546
Chemical and Rubber Products	694	1,344	2,201	2,354	5,327
Petroleum and Coal Products	—	—	2	83	—
Glass and Stone	24	90	729	1,931	2,752
Steel and Metal Products	1,639	1,434	12,514	7,965	17,867
Machinery (including electrical)	785	406	1,842	1,341	3,804
Transport Equipment	150	1,042	2,242	853	1,629
Miscellaneous Manufacturing	1,718	1,402	2,787	5,021	8,791
TOTAL	38,648	54,804	86,796	118,860	174,998

Source: Wontack Hong, *Factor Supply,* Table A-12

Note: Total includes "other services," scrap iron, and "unclassifiable" exports.

By contrast, the three largest export sectors in the late 1950s and early 1960s had been Other Agriculture, Processed Food and Beverages, and Minerals. Although exports grew even for these predominantly primary-based industries, their share fell from 67 percent of exports as late as 1961 to 28 percent by 1965. This shift in the composition of commodity exports is further reflected in the data in Table 26. The drop in the relative

TABLE 26 Percentage Distribution of Exports, 1960–1965

Sector	*Average* *1950 to 1960*	*1961*	*1962*	*1963*	*1964*	*1965*
Agriculture	19.9	14.4	21.8	6.4	6.2	5.2
Minerals	53.0	38.3	22.3	17.5	16.8	13.6
Total Primary	75.1	57.0	52.9	30.6	27.7	23.6
Manufactures	24.9	43.0	47.1	69.4	72.3	76.4

Sources: Data for the 1950 to 1960 average are from Table 13. Percentages for 1961 to 1965 are derived from Table 25.

Note: Total Primary includes forestry and fisheries, which are not listed separately.

importance of primary-commodity exports and the increase in that of manufacturing exports was extremely abrupt. Mineral exports, which had accounted for more than half of export earnings in the 1950s, declined from an average of 53 percent of total exports in the 1950s to less than 14 percent by 1965, despite the fact that, in absolute terms, earnings had risen from $14.8 million in 1961 to $23.7 million in 1965. Indeed, the share of agriculture and minerals fell even more than the total primary percentage indicates, as exports from forestry and fisheries, primarily the latter, rose sharply (see Table 13). The net result was that, by 1965, over three-quarters of Korean export earnings originated in manufacturing. To be sure, in value-added terms, the contribution of manufactures was somewhat less, as most manufactured exports had a relatively high import content. This consideration is dealt with in greater detail in Chapter 4, but it does not fundamentally alter the basic conclusion that the relative importance of manufacturing as a source of foreign exchange earnings increased markedly during the early 1960s, and that exports grew over a wide range of manufacturing sectors. In the late 1950s, Korea had been a net exporter of primary commodities and a net importer of manufactures. This balance was changing throughout the process of rapid export growth.

The geographic destination of exports is also of interest from the viewpoint of analyzing the reasons for Korea's successful transition to export promotion. Some observers of world trade patterns have argued that the phenomenal Korean success is essentially non-repeatable because it was Korea's close ties with, and proximity to, Japan that enabled the rapid growth of exports during the 1960s. This argument does not bear close inspection for two reasons: first, trade relations between Japan and Korea were not normalized until the mid-1960s; second, the period of rapid export expansion was accompanied by a decline in the relative importance of Japan as a destination for Korean exports. The total dollar value of exports, and percent of total exports, going to Japan was as follows: [15]

TABLE 27 Total Dollar Value of Exports, and Percentage of
Total Exports, to Japan, 1960–1965

	Exports to Japan ($ millions)	% of Total Commodity Exports
1960	19.6	61.5
1961	18.3	47.5
1962	23.5	42.8
1963	24.8	28.6
1964	38.1	32.1
1965	37.6	25.1

As can be seen in Table 27, Japan's imports from Korea did not even double during the period when Korea's exports rose six-fold. Reflecting this, the Japanese share, which had been 61.5 percent in 1960 (and had averaged 53.5 percent during the five-year period 1955 to 1960), fell sharply between 1960 and 1965. The two geographic areas that absorbed an increasing share of Korea's exports during the 1960–1965 period were the United States (whose share rose from 11.1 percent in 1960 to 35.2 percent in 1965), and East Asia other than Japan (whose share rose from 11.3 percent in 1960 to 22.8 percent in 1965).

It seems evident, therefore, that rapid Japanese economic

growth did not contribute a rapidly growing market that can explain the rapid growth of Korean exports. Instead, Korean export growth was significantly greater in other markets than in Japan which, during the crucial years of transition to an export-oriented trade policy, expanded her imports from Korea relatively slowly.

It is not possible to quantify the contribution of the export-promotion policies described in the last section to the success of the transition to an export-oriented economy in the early 1960s. It seems evident that without those incentives, or at least a significant shift in incentives from those that had prevailed in the 1950s, the very rapid growth of export earnings could not have been realized and sustained. The government's commitment to the export-promotion policy, and its willingness to adjust incentives in response to the behavior of exports, must have been a powerful contributing factor, in that it provided assurance for those contemplating entering export markets that, if they successfully competed abroad, profitability would continue.

It also seems apparent that one cannot attribute Korea's success entirely to luck. World markets were growing during the 1960s, and it was easier for the Korean export-promotion policies to have the desired effect against the background of growing world markets that it would have been if the world economy had been stagnant. Other countries, however, were confronted with similar world market conditions and did not achieve anything like the Korean results. Whether Korean policies would have been successful against the background of international recession is arguable; it does not, however, seem to be possible to make the case that fortuitous external events explain Korea's successful export performance.[16]

IMPORTS

As seen earlier, the 1961 devaluation was followed by a short-lived liberalization of imports and then a reversion to greater reliance upon quantitative controls, while the 1964 devaluation was followed by continuing liberalization. During the transition years as a whole, therefore, quantitative restrictions continued to play an important role in determining the commodity composition of imports, although their relative importance diminished fairly rapidly toward the end of the period.

Table 28 gives the total dollar value of imports for each year from 1961 to 1965, and also the breakdown of imports by major categories in both dollar and percentage terms. The import boom that accompanied domestic inflation and occasioned the reimposition and intensification of quantitative restrictions in 1963 is perhaps the most prominent feature of the statistics.[17] Also to be noted is the sharp reduction in imports in 1964, the combined result of fairly tight import restrictions early in the year and the devaluation and accompanying special tariffs later in the year.

By and large, the commodity composition of imports appears to have changed relatively little between the latter half of the 1950s and the first half of the 1960s. To be sure, comparison is somewhat difficult in the presence of a sizable proportion of "unclassifiable" imports in the late 1950s. Even so, it would appear that imports of finished consumer manufactures had probably been curtailed to the extent deemed feasible by the mid-1950s, and that the breakdown of imports among raw materials, intermediate goods, and investment goods was much the same, regardless of the changed orientation of the economy, in the two periods.

Despite the shift in emphasis toward exporting, and the apparent stability of the structure of imports among end-use categories, import substitution continued in a number of industries in the early 1960s. In contrast to the widespread expansion of exports, the import-substitution thrust was relatively more concentrated, and less across-the-board in nature, than in the 1950s. Suk Tai Suh's data show sharp drops in the percentage of domestic demand satisfied by imports in paper and paper products (from 30 percent supplied by imports in 1960 to 12.5 percent in 1965), coal and petroleum products

TABLE 28 Commodity Composition of Imports, 1956–1960 Average, and 1961–1965 ($ millions—figures in parentheses indicate percentages of total imports)

	Food and Beverages	Crude Materials	Mineral Fuels	Chemicals	Manufactures	Machinery & Transport Equipment	Total
1956–1960 average	56.7 (15.3)	64.5 (17.4)	37.5 (10.1)	73.8 (19.9)	61.9 (16.7)	41.5 (11.2)	370.8
1961	40.1 (12.7)	67.3 (21.3)	27.5 (8.7)	61.6 (19.5)	45.2 (14.3)	42.4 (13.4)	316.1
1962	48.9 (11.6)	93.6 (22.2)	30.8 (7.3)	94.5 (22.4)	83.5 (19.8)	69.6 (16.5)	421.8
1963	121.0 (21.6)	112.0 (20.0)	34.2 (6.1)	80.1 (14.3)	96.4 (17.2)	115.4 (17.2)	560.3
1964	68.3 (16.9)	101.1 (25.0)	28.3 (7.0)	84.5 (20.9)	51.4 (12.7)	69.6 (17.2)	404.4
1965	63.5 (13.7)	114.0 (24.6)	31.5 (6.8)	103.3 (22.3)	77.4 (16.7)	73.7 (15.9)	463.4

Source: BOK, *Economic Statistics Yearbook.*

Note: For 1956 to 1960, an average of 9.4% of imports were placed in an "unclassifiable" category. In 1961, 10.1% were put in that category. After 1961, less than 1% were so classified. For that reason, the dollar figures do not sum to total imports, and the percentages do not add to 100.

(from 34 percent in 1960 to 3.4 percent in 1965),[18] chemical fertilizers (where 100 percent of domestic demand had been supplied by imports until 1960, and one-fourth was met by domestic production by 1965), and electrical machinery (from 45-50 percent imported in 1960-1961 to 18 percent imported in 1965).[19] In some cases, such as chemical fertilizers, these figures represent in part the gestation lag from the time of import-substitution investments in the late 1950s until output was realized some time later. By and large, however, the data correctly reflect the fact that the early 1960s was a time of transition: selective import substitution continued to take place simultaneously with the expansion of production for export.

There was, nonetheless, a sharp reversal in the relative contribution of export expansion and import substitution to growth. It will be recalled that Frank, Kim, and Westphal estimate that the total contribution of export expansion to growth was 12.9 percent from 1955-1960, while the direct contribution to manufacturing growth was 5.1 percent. By comparison, import substitution contributed 10.2 percent total, and 24.2 percent directly to manufacturing growth performance. Import substitution had contributed about five times as much in manufacturing, and almost as much as export expansion in total. The Frank, Kim, and Westphal estimates for ensuing periods are divided into the subintervals 1960-1963 and 1963-1966. Their estimates of the respective contributions are shown in Table 29.[20] Import substitution's overall contribution was negative during the period 1960-1963, and its contribution to manufacturing

TABLE 29 Contributions of Export Expansion and Import Substitution to Growth, 1960-1963, 1963-1966

	1960-1963		1963-1966	
	Direct Manufacturing	*Total*	*Direct Manfacturing*	*Total*
Export Expansion	6.2	6.3	29.4	31.4
Import Substitution	-0.9	- 6.9	14.4	8.9

growth was smaller in absolute value, but also negative. After 1963, it would appear that import substitution again contributed positively to growth, albeit by a far smaller percentage than export expansion.

CAPITAL FLOWS AND AID

It was seen in Chapter 2 that aid had begun to diminish in the late 1950s. That trend continued into the early 1960s. Current account deficits persisted throughout the transition to export-oriented growth. Moreover, it was anticipated that deficits would continue. Since aid was expected to decline in absolute importance, the government began adopting a series of measures to encourage the inflow of foreign capital.

The first foreign loans, other than those financed by the American Development Loan Fund, were negotiated during the early 1960s, under the newly passed laws for encouraging foreign capital. By and large, however, private foreign capital remained a relatively small source of foreign exchange earnings during the 1961–1965 period. As with exporting, the early 1960s was a time of transition as policy shifted toward the inducement of foreign capital. Unlike exporting, however, capital flows did not begin increasing significantly until after the transition had been completed.

Table 30 provides data derived from the balance-of-payments and national-income accounts which yield some indication of the relative importance of aid, and of other sources of foreign exchange, in financing real resource accumulation during the 1960–1965 period. Imports are recorded f.o.b., and therefore do not correspond with the data found elsewhere in this volume. As can be seen, private capital flows were negligible, and cumulatively negative, through 1962. Thereafter, they were positive but still relatively small. Until 1961, net foreign assistance was of approximately the same order of magnitude as the current account deficit, and covered three-quarters or more of

TABLE 30 Current Account Balances, Private Capital Flows, and Aid, 1960–1965 ($ millions)

	1960	1961	1962	1963	1964	1965
Imports f.o.b.	306	283	390	497	365	416
Current Account Balance	-262	-198	-292	-403	-221	-194
Private Capital Flows	3	-2	-4	61	7	17
Total Net Foreign Assistance	256	207	200	208	141	134
Aid as % of Imports	83.6	73.1	51.2	41.8	38.6	32.2
Imports as a % of GNP	12.6	14.9	17.1	16.4	13.9	16.0
Current Account Deficit as a % of GNP	9.3	9.5	11.9	11.5	7.8	7.4

Sources: IMF, *International Financial Statistics*, (August 1976), and BOK, *Economic Statistics Yearbook, 1976.*

Note: Computations of the relative importance of imports, aid, and the deficit were made in the way described in Table 18.

imports. Thereafter, aid began diminishing rapidly as a proportion of foreign-exchange resources, as exports began growing. By 1965, imports were over $400 million (even on an f.o.b. basis), and accounted for 16 percent of GNP. Aid, however, financed only 32.2 percent of the import flow in that year, as exports had risen from about one-tenth of imports to over two-fifths of imports.

It should be emphasized that the transition to an export orientation was accomplished with an increase in the relative importance of imports as a fraction of GNP. The move to an outward-looking strategy implied changing the structure of the economy in such a way that *both* exports and imports increased in relative importance.

Equally important from the viewpoint of understanding the transition and the role of the foreign sector in Korea's modernization is the fact that the current account deficit, as a percentage of GNP, remained sizable even until the late 1960s. Foreign resources continued to be important as a source of savings,[21] although those resources originated predominantly from private foreign sources, especially after 1965.

ENCOURAGING PRIVATE FOREIGN CAPITAL

As already indicated, foreign-capital inflows did not become quantitatively important until 1966. In part, this was because of long lags between the time when foreign-loan agreements were approved and the time when resources began to be realized from those loans. In part, however, it was simply because the policies that induced the inflow of foreign capital required time to be developed, and initial successes with exporting were necessary to alter foreigners' expectations as to Korean prospects.

As early as January 1960, a Foreign Capital Inducement Law was promulgated. Prior to that date, the only foreign loans had originated from the AID Development Loan Fund, and total loan arrivals through 1961 totaled only $4.7 million.[22] In 1962, the government identified nine major projects included in the Five-Year Plan that would require foreign capital, and sent an economic mission abroad to attempt to secure financing for the projects. In that same year it supplemented the original law, enacting provisions whereby imports could be financed by long-term export credits and also providing repayment guarantees. In addition, tax concessions were granted as a further inducement to foreign capital.

Frank, Kim, and Westphal summarized the results:

> Because of the positive measures of the government to attract foreign capital, foreign loans and investments "finalized" increased sharply after 1962 and amounted to $222.7 million at the end of 1963 . . . Foreign loans finalized at the end of 1960 were only about $18.8 million. At the end of 1963, commercial loans finalized amounted to $127.5 million, larger than the $84.4 million of finalized foreign public loans. Actual "arrivals" of the foreign loans and equity investment were, however, relatively small in 1961–63 . . . since finalized foreign loans and investment generally required a year or more before the goods and services financed by the foreign capital actually arrived.[23]

The next measures were not taken until 1966. That year began the major inflow of foreign capital which so clearly demarcates the late 1960s, when foreign capital financed the current account deficit, from the early 1960s, the last years in which aid dominated the financing of the current account deficit and of imports. Further discussion can therefore be postponed until Chapter 4.

AID DURING THE EARLY 1960s

In most respects, aid during the early 1960s followed a course similar to the pattern set in the 1950s. It continued to originate exclusively from the United States and was predominantly in the form of grants rather than loans. Differences were twofold: 1) there appears to have been considerably less friction between donor and recipient, partially as a result of the stabilization program and subsequent policy changes and partially because of the change in governments; and 2) whereas the Korean economy and economic policy were heavily dependent upon a sustained flow of aid in the 1950s, the level of aid was already considerably reduced from its 1957 high; and donor-recipient relations were predicated on the assumption that the phase-out of aid would continue. It was already mentioned that acceptance of this prospect was a crucial factor in leading to the decision to embark upon the export-promotion strategy.

Table 31 provides data on the total aid flow and its breakdown into major components from 1961 to 1965. The data are comparable with those given in Table 18. It will be recalled from that table that foreign assistance had reached $383 million in 1957 and had already fallen to $245 million by 1960. As is apparent from Table 31, aid inflows remained at approximately that level through 1963, and then once again fell sharply to $165 million in 1964, remaining at about that level in 1965. Non-project assistance continued to be the major form of support, although PL 480 aid increased in both absolute and proportionate importance in the early 1960s. In fact, in 1964, PL 480 sales alone exceeded the total of supporting assistance.

Korea continued to be virtually unique among the developing countries in that the preponderance of aid was received in the form of grants rather than loans. There had been no loans received prior to 1959. Thereafter, some aid was channeled through the Development Loan Fund. However, by the end of 1960, total receipts under DLF were on the order of $1.3 million, less than 1 percent of aid in either year. As can be seen, they were somewhat larger in the early 1960s, although the fraction of non-grant assistance remained very small. It was not until 1965 that the United States committed itself to long-term loans as a means of continuing support.[24]

An indication of the types of activities supported by aid can be gained by inspecting the sectoral distribution of non-project and project assistance. In a 27-sector breakdown, chemical fertilizers were the largest single sector for non-project supporting assistance in the late 1950s and in the first half of the 1960s, with an average annual support level of $43.2 million for 1956–1960 and $33.6 million in 1961–1965. Support for petroleum

TABLE 31 Total Aid Received, by Source, 1961–1965
($ millions)

	1961	*1962*	*1963*	*1964*	*1965*
Non-Project Supporting Assistance	113.6	126.6	102.7	72.8	79.2
Project Assistance	29.8	21.7	13.0	5.5	4.3
PL 480 Sales	32.6	36.1	62.7	94.7	54.4
PL 480 Title II and III	10.2	24.0	21.8	27.6	28.5
Development Loans	3.2	10.5	20.0	4.5	2.6
TOTAL	192.8	245.5	252.3	164.8	176.9

Source: Data provided by USAID, Korea.

Note: Data represent financial expenditures on a calendar-year basis. Data do not correspond to balance-of-payments figures given in Table 30 due to differences in timing, valuations, and concepts. For example, USAID payroll in Korea is included in this table, but not in Table 30.

products was also sizable until 1963, averaging $20.2 million annually between 1955 and 1960, and $24.7 million between 1961 and 1963. The support level was $8 million in 1964, the last year for which there was non-project assistance to that sector. Assistance to rice, barley, and wheat, to other agriculture, and to fiber spinning was sizable in the late 1950s but diminished sharply in the early 1960s. By contrast, project assistance was much more heavily concentrated. In the late 1950s, public utilities and other construction accounted for over three-fifths of all project-commodity assistance. The only other sector to receive a significant share was transportation equipment, but that had largely ceased by the early 1960s.[25]

Counterpart funds (see pp. 74–75) continued to be used to provide the domestic component of financing investment projects, in much the same way as had occurred in the late 1950s. As with aid patterns in general, the only significant change was that domestic savings were increasing and, because of that, decisions with respect to counterpart funds were relatively less important in the allocation of investment resources than had been the case earlier.

As with so many phenomena, then, aid in the early 1960s was in a state of transition. It continued to be important, accounting for over half the import bill and a very high fraction of investible resources. Nonetheless, its importance was diminishing, and the period was essentially one during which aid moved from center stage. By the time the transition to an export-oriented economy was completed, the Korean government had also succeeded in attracting sizable foreign capital, which to a large extent substituted for aid in maintaining a high fraction of foreign savings in the total.

REASONS FOR SUCCESS OF THE TRANSITION

By 1965, the success of the export-promotion policies in achieving rapid export growth was evident. Because that success was associated with a sharp increase in the rate of economic growth, the export orientation of the Korean economy became an accepted, and largely unquestioned, basis for policy. There was no longer any doubt as to whether export growth should outpace GNP growth: the only question was by how much it should do so.

A number of questions about the export-promotion strategy are of considerable interest. Among the more important are 1) the extent to which export-promotion policies were themselves responsible for, or contributed to, the more rapid rate of growth and 2) the efficiency of the export-promotion policies. Those questions must await analysis of the period of sustained and rapid growth of exports and real output from 1965–1972. At this stage, however, there is a third important question that can be addressed. This is the analysis of why Korea was so successful in changing her trade orientation. Many countries have, after all, embarked upon an export-promotion campaign at one or more times. Often, exports respond to some extent with a sharp spurt, only to fall off again. When that happens, governments frequently are forced to retreat from their determination to achieve rapid export growth. In fact, in most countries, the strategy of import substitution has been adopted largely because of pessimism with regard to the possibility of achieving sustained growth of exports. Why was it, then, that Korea could embark upon such a strategy and somehow achieve sufficient success so that the policy became self-reinforcing?

A number of factors contributed. First, and of considerable importance, was the fact that Korea's exports had languished during the 1950s. By 1960, at the time the strategy switch occurred, therefore, Korea's exports were abnormally low relative to her size and stage of development. In that sense, there was something of an "export potential" waiting to be tapped. That potential provided a reservoir from which initial spurts in exports could occur in response to the incentives offered to them. After all, it did not require a very large increase in the absolute level of exports to result in a very sizable percentage increase, given the initially low base.

The fact that there was some backlog of export potential, however, was by no means sufficient to account for the initial success of the export-promotion strategy. In particular, that potential could have remained totally unrealized in the absence of a set of incentives that made exporting a profitable activity, on a par with alternative activities that private-sector entrepreneurs might undertake. A large part of the initial success of the Korean export-promotion drive must be attributed to the fact that the incentives for export, both explicit (in the form of real exchange rates, export subsidies, and the other measures listed in Table 24) and implicit (in the form of expedited government action and preferential treatment for exporters) were sufficiently strong and sustained. Perhaps the most difficult challenge that most governments face in attempting to reverse earlier inward-looking policies is that of convincing would-be exporters that the commitment to the export strategy and incentives for exporting will continue. In the Korean case, the government's commitment was exceptionally strong, and it was not only the fact of incentives, but the government's willingness to alter them to induce the desired performance that was undoubtedly important in bringing about rapid export growth in the early 1960s.

In that regard, an important piece of evidence is that the 1961 devaluation was not notably successful. Indeed, the real exchange rate and the value of export incentives began dropping sharply in 1962. Simultaneously, the tightening of quantitative restrictions must have increased the profitability of import-substitution activities. However, the government reacted by providing export subsidies and other inducements to exporters which, it was shown, resulted in an increase in the real proceeds per dollar of exports despite the appreciation of the real exchange rate.

The 1964 devaluation, and commitment to maintaining the real value of the exchange rate by adoption of the sliding peg, was undoubtedly a precondition for the export growth of the late 1960s. Accompanying the devaluation were measures aimed at financial reform[26] which were necessary if Korea was to be able to induce private capital flows of any magnitude.[27] Without the 1964 reforms, it is doubtful whether continued high rates of export expansion could have been realized. Had the success prior to that date been less, it is possible that the government would not have had the political support and determination to carry out the reforms necessary to transform the initial, tentative, success with exporting into the sustained drive it became in the late 1960s.

The appropriate conclusion, therefore, would appear to be that the initial set of export incentives, combined with the low export base and the untapped potential that then existed, accounted for the rapid growth of exports in the 1961–1964 period. While there were twists and turns in policy during the 1961–1964 years, these were more the unintended result of other policies (such as the impact of inflation on the balance of payments) than of any retreat from the export-promotion strategy. Export growth, in turn, reinforced the commitment to an export orientation, and did so in a way that enabled the devaluation and other reforms of 1964–1965. Thereafter, the government was able to maintain a continuity of policy which was undoubtedly necessary for the export expansion of the next decade.

FOUR

Emergence as a Major Exporter, 1966 to 1975

Beginning in 1966, the policies that had been established during the transition years had their payoff: export growth was rapid and sustained, Korean exporters became established in international markets in a variety of lines, and Korea emerged as a major competitor in international markets.

From any long-term historical perspective, the decade from 1966 to 1975 will undoubtedly be viewed as homogeneous with regard to the trade-and-payments regime: the commitment to an export-oriented strategy remained unchallenged; the policy instruments employed in pursuit of that strategy were fairly stable with only minor and gradual changes; and the underlying trend in export and import growth remained much the same throughout.

Viewed from the closer perspective of 1976, however, there were some shifts and changes that demarcate sub-intervals of the decade. For that reason, it is convenient to begin with a brief outline of the chronology of those events affecting trade and payments. Thereafter, the trade-and-payments regime, and the behavior of exports, imports, capital flows, and foreign aid can be discussed in turn.

AN OVERVIEW OF THE PERIOD

Any attempt to divide Korea's trade-and-payments history into well-defined sub-intervals is bound to have an arbitrary element, but the most questionable division is any demarcation of the post-1960 period. Kwang Suk Kim, for example, treats 1960 to 1963 as the transition period, and regards the period from 1964 onward (through 1973, the end of the period he covered when writing in 1974) as years of rapid export growth.[1] Frank, Kim, and Westphal, by contrast, treat 1961 to 1966 as the transition period; they treat the years 1967 to 1972—the end of the time period they covered—as the years of sustained export growth.[2] The reasons for the difficulty in placing the end of the transition and the start of sustained growth are not difficult to pinpoint: as shown in Chapter 3, policy instruments were frequently altered during the switch to export promotion in response to the degree of success in achieving rapid export growth, and it was not until 1967 that the last export-incentive measures were introduced (see Table 24). One can, therefore, make an excellent case for dating the end of transition as the time when export growth was really sustained, which would be Kim's cutoff, and an equally valid case for dating the end of the transition as the time when stability in incentives had been achieved—either 1966 or 1967.

Once that demarcation is made, there is little doubt that the remainder of the period was characterized by rapid export growth, as exports rose from $250 million in 1966 to $835 million in 1970, $1,624 million in 1972, and $5,081 million in 1975. This rapid growth was reflected in Korea's rapidly increasing share of world exports, which was less than 3/100ths of 1 percent in 1960, rose to 14/100ths of 1 percent in 1966, 0.29 percent in 1970, 0.43 percent in 1972, and 0.61 percent in 1975.[3]

Within the period, however, there were slight changes in emphasis. These came about primarily as a result of the immediate balance of payments position and prospects. Until 1973, changes were relatively minor and designed to offset shifts in the net foreign-exchange position. After 1973, the sharp changes in the international economy impinged upon the payments position and influenced the overall direction of policy toward the trade sector.

By and large, one can characterize the changes as fluctuations in the extent of liberalization of the trade-and-payments regime. In 1967, after it became clear that rapid export growth was a reality, an effort to liberalize the import regime was made. This effort had several components, including the shift from a positive list (of permitted imports) to a negative list (of prohibited imports), which left a long-term imprint on the regime and resulted in sustained liberalization. There was also an effort to reform the tariff structure, but overall tariff reduction was not achieved.

After the middle of 1968, efforts comparable to the 1967 liberalization were no longer made and, indeed, the regime turned somewhat more restrictive in response to payments pressures. Borrowing from abroad was increasing rapidly in the period: as of 1965, accumulated debt and debt-servicing obligations were simply not a factor with which the authorities had to cope. By the late 1960s, debt-service obligations contributed as much, if not more, to concern about the payments position as did the trade balance, and the stringency of the regime altered in response to that. These shifts in emphasis—albeit relatively minor—were reflected in the stated objectives of the foreign-exchange budgets and trade programs announced each year. For example, the 1970 budget placed emphasis upon curtailing imports of "nonessential goods" and increasing support for export industries,[4] and the 1971 and 1972 budget further emphasized restrictions upon imports.[5] By 1973, the "general principles" were somewhat altered: 1) emphasis was to be placed upon increasing the capacity to repay foreign debts; 2) higher "priority" was to be given to export industries; and 3) restrictions upon imports were not to obstruct the "efficient supply of raw materials and of goods required for stable economic growth."[6] For 1974, top priority was given to raw materials procurement, including assuring availability of foreign-exchange loans for importers,[7] while for 1975 emphasis once again shifted to restraining imports.

With each of these shifts, various categories of imports were shifted from AA to restricted status, and conversely. Import deposit requirements and other measures were simultaneously adjusted, and numerous other relatively minor measures were undertaken. None of them fundamentally altered the export orientation of the economy, however, and export growth remained the fundamental commitment of the government.

Insofar as any change in that commitment can be discerned, it came relatively late in the period, and seems to have originated from uneasiness stemming from the rapid world inflation and increase in the price of oil in 1973–1974.[8] At that time, the first questioning of the extent of the commitment to export was heard. To be sure, there were no suggestions that the economy should reverse its fundamental orientation. Rather, there was a great deal of discussion of "dependence" on foreigners, and suggestions were voiced to increase the reliance of exporters upon domestically produced intermediate and capital goods. As the preparations for the Fourth Five-Year Plan got under way, debate over the export target reflected this concern. By mid-1975, the Korean economy had demonstrated its resilience in the face of the oil price increase. By 1976, the policy questions implicit in the debate seem to have been resolved without a noticeable swing toward encouragement of production for domestic consumption. Nonetheless, the debate reflected the first time that the extent of emphasis on export promotion was at all seriously questioned in public discussion. Even then, it should be noted, there were no advocates of a significant reversal of policy.

Thus, if one were to attempt to identify sub-periods within the 1966–1975 period, the early years would be characterized as attempts at further liberalization of the regime within the context of rapid export growth. The middle years, from 1968 to 1972, were notable for the emergence of debt-service obligations as a significant component of the balance of payments and also for the failure of the Korean government to liberalize its import regime further despite the rapid growth in foreign exchange resources. The final years were marked by the upheavals of the international economy and the successful use of policy instruments to enable the Korean economy to adapt remarkably well.

THE TRADE–AND–PAYMENTS REGIME

Table 32 gives the basic data on nominal, effective, real, and purchasing-power-parity exchange rates for the 1966-1975 period. Despite much greater stability in real rates than had characterized earlier periods, the precise mechanism by which exchange rates were determined altered on several occasions.

EXCHANGE RATE POLICY

It will be recalled that the government had implemented a floating unified exchange-rate policy in March 1965. At that time, the rate was 270 wŏn per dollar. It actually declined to 256 wŏn per dollar by the end of April, and then rose to 280 wŏn per dollar by the end of May. Starting in June, the Bank of Korea began undertaking limited intervention in the foreign exchange market. By August, the Bank was selling exchange certificates at 271 per dollar. That completely repegged the exchange rate, which remained at that level throughout 1967. In 1968, intervention policy was again altered: the wŏn was permitted to depreciate slowly in an amount deemed sufficient

TABLE 32 Nominal, Effective, and Purchasing-Power-Parity Exchange Rates for Exports and Imports, 1966–1975

	1966	1967	1968	1969	1970	1971	1972	1973	1974	1975
A. Official Exchange Rate (wŏn per dollar)	271	271	277	288	311	348	392	398	407	485
B. Export Subsidies (wŏn per dollar)	52	62	78	80	88	103	105	94	86	81
(Internal tax exemptions)	(21)	(23)	(23)	(31)	(30)	(37)	(28)	(22)	(22)	(34)
(Customs duty exemptions)	(21)	(25)	(40)	(34)	(40)	(48)	(66)	(64)	(55)	(34)
(Interest rate subsidies)	(10)	(15)	(15)	(15)	(17)	(18)	(11)	(7)	(9)	(13)
C. Export EER (A + B)	323	333	355	368	399	451	497	493	493	566
D. PLD EER for Exports (C divided by 1965 price level)	297	286	283	275	273	284	275	255	180	163
E. PPP PLD EER for Exports (D times index of price level of major trading partners adjusted for yen revaluation)	305	297	299	299	308	325	349	396	338	321
F. Actual Tariff Equivalents (wŏn per dollar)	25	26	26	25	26	22	23	19	19	25

TABLE 32 (continued)

	1966	1967	1968	1969	1970	1971	1972	1973	1974	1975
G. EER for Imports (A + F)	296	296	303	313	336	369	415	418	425	510
H. PLD EER for Imports	272	256	242	234	231	233	230	216	155	147
I. PPP PLD EER for Imports	280	266	255	255	260	270	290	332	288	287

Source: Larry E. Westphal and Kwang Suk Kim, "Industrial Policy and Development in Korea," (mimeo, World Bank Staff Working Paper No. 263, August 1977), Table B.

Note: The value of railroad and electricity discount is included in total export subsidies for 1971 and 1972. See Appendix A for an explanation of exchange-rate concepts.

to maintain purchasing-power parity (that is, the wŏn depreciated by the weighted percentage difference in inflation rates between Korea and her major trading partners). That policy continued until June 1971, when there was an abrupt 13 percent devaluation from 326 wŏn per dollar to 370 wŏn per dollar. The rate remained pegged at that level until the end of 1971, then was allowed to depreciate until June 1972 when it was again pegged, this time at 400 wŏn per dollar. When the American dollar was devalued in February 1973, it was decided to maintain the wŏn-dollar rate. That decision, of course, represented a sizable devaluation relative to Japan. The 400-wŏn-per-dollar rate remained in effect until December 1974, when it increased to 484 wŏn per dollar, the rate that prevailed until the end of 1975.

In real terms, the decade really consists of two periods. The first lasted from 1966 to 1973. During that time, the export PLD EER was maintained even while the exchange rate was pegged by altering the value of export subsidies. The fluctuation in the PLD EER for exports during the years 1966–1971 was less than 2 percent. Since world prices and exchange rates were fairly stable, the PPP PLD EER did not alter much, although world inflation after 1969 meant that the policy of holding the PLD EER virtually constant tended to increase the competitiveness of Korea's exports. The treatment of imports stood in sharp contrast to that of exports. When the nominal exchange rate was constant, the real exchange rate for imports appreciated; when that happened, the government took measures to increase the restrictiveness of quantitative controls. Up until 1972, therefore, the exchange-rate regime was asymmetric: on the export side, it was recognized that the real return for exporting had to be maintained at a fairly realistic and constant level. Pricing incentives were therefore used to supplement the exchange rate. On the import side, the price of imports was not relied upon as the only or even the major means of allocating foreign exchange. Quantitative restrictions played a larger or smaller role, depending upon the foreign-exchange situation.[9]

After the dollar devaluation in February 1973, the PPP PLD EER rose sharply to 396, compared to 308 wŏn per dollar in 1970. By 1974, it was 338 1970 wŏn per U.S. dollar, and in 1975 it was further adjusted to 321. This realignment was in part effected by failing to increase export incentives and in part by failing to alter the exchange rate promptly as Korean inflation exceeded that in the rest of the world.

The export incentives, which served to maintain the real export exchange rate at times when the nominal rate was stable, are given in row B of Table 32. They fail fully to reflect the value of all incentives, but provide an indication of the major ones. As can be seen, customs-duty exemptions were quantitatively the most important of the three, and interest-rate subsidies were the least important. Direct tax preferences for exporters (but not indirect tax exemptions) were abolished in 1973 as were automatic tariff exemptions for exporters on imported capital equipment. These changes reflected in part the effort to offset the increase in the PPP PLD EER which had come about over the preceeding several years.

Even the customs-duty exemptions were altered after 1973: prior to that date, exporters had been exempt from the duties; in 1973, the Korean government decided to shift to a drawback, or rebate, system.[10] The initial effect of this shift was not felt until 1975, however, and even then its impact was softened because exporters were given the right to postpone payment of duties for specified periods. It was announced that the grace period would be gradually reduced, and finally ended in 1979. By 1975, internal tax exemptions were as important as customs exemptions as an export incentive. Altogether, the value of the quantifiable subsidies exceeded 20 percent of the official exchange rate throughout the period, and was almost one-third of the official rate in 1971. If the value of all the other export incentives could be quantified, the value of export incentives would appear even greater. As explained by Cole and Lyman:

While the weight of incentives, both direct and indirect, shifted increasingly away from import substitutes and toward exports in 1965 and 1966, a second major factor was also working in the same direction. This was the political and administrative backing for an all-out government campaign to expand exports, typified by the constant setting of seemingly unattainable export targets, their attainment, and then the setting of even higher targets. The president took a strong personal interest in export expansion and was primarily responsible for continuously elevating the targets. He held monthly meetings to review the progress of the export drive and to ensure that no administrative obstacles impeded export growth. Procedures were simplified; special consideration was given to exporters who were having difficulty filling their orders; and embassy staffs abroad, up to and including the ambassadors, were pressed into service as export-promoters . . . The political leadership made it clear that performance would be judged on what an individual or agency had contributed to the growth of exports.[11]

By 1969, exporters were being graded into four classes on the basis of their performance, and the National Medal of Honor was awarded to the highest achievers. Moreover, tax surveillance of the outstanding performers was deliberately relaxed as a matter of policy.[12]

For all these reasons, the relative inducement to export, compared with import substitution, was probably even greater than the ratio of the export to the import EER that Table 32 indicates. To be sure, there were some quantitative restrictions on imports of commodities for which import-substitution policies had been adopted, and the implicit value of that protection was undoubtedly very great for some domestic industries. However, as indicated above, import-substitution policies were carried out selectively, so that the weights attached to the value of the omitted QR-induced incentives would be relatively smaller than the weight associated with the additional export incentives. Even without taking into account the incentives whose value could not be quantified in Table 32, the export EER exceeded the import EER by 8 percent in 1966. Thereafter, the gap widened to about 17 percent in 1968, and reached about 25 percent in 1971, narrowing somewhat again after the devaluation of 1972. Even then, it remained at 11 percent in 1975, even when calculations are based solely on the export incentives enumerated in Table 32.[13]

EXCHANGE CONTROL

Despite oft-repeated statements of intent to liberalize, reliance on quantitative restrictions for governing foreign-exchange expenditures fluctuated over the 1966–1975 period, with at least a small trend toward increasing liberalization. The only significant and lasting move in the direction of liberalization was the shift from a positive-list to a negative-list system for controlling imports.

If one had a complete enumeration of all possible commodities that might be imported, it would make little difference whether an import regime was based on a positive list—itemizing all the items for which approval to import would be granted—or a negative list under which permission would be granted unless the item were specifically listed as prohibited. In practice, there are so many commodities that complete itemization is impossible, and the distinction can be quite important: under a positive-list system, import licenses are granted automatically only when an authorized official can find the item specifically listed on the approval list; under a negative-list system, an official grants the license unless he finds the item on the negative list. A negative-list system is therefore considerably less restrictive than a positive-list system.

The Korean government shifted from a positive- to a negative-list system in July 1967. This makes comparison of the number of items on each list prior to and after that date meaningless,[14] since the shift itself represented a considerable liberalization of the import control system. After that date, commodities were shifted between lists, both in response to the domestic supply-demand situation and in reaction to the degree to which the foreign exchange situation was perceived to be comfortable. An indication of those trends can be gleaned from enumeration of the items on each list. These data are given in Table 33. When the shift to a negative-list system was made, it was intended as the beginning of a major liberalization effort. Indeed, as can be seen, this intent was carried out through 1968, as the number of prohibited items declined from 118 to 71. Thereafter, however, changes were effected primarily by shifting commodities between the restricted and the AA list. The number of restricted commodities increased in 1969, fell a little in 1970, rose again in 1971 and 1972, declined in 1973, and increased in 1974 and 1975. These data, despite their shortcomings, fairly accurately reflect the underlying trend in the regime: there were short-term shifts between more and less restrictionist content of the import regime, but there was little underlying long-term trend.

The same pattern was reflected in other aspects of control. All invisible transactions were licensed, and the amount of foreign exchange that could be purchased for various categories of transactions changed from time to time. Guarantee deposit requirements, likewise, were altered in light of the authorities' anticipations of import demands and foreign exchange availability.

An enumeration of some of the changes made during the last half of 1968 illustrates both the variety of control instruments and the manner in which they were altered. As already indicated, the second half of 1968 was a period when concerns were being expressed about foreign exchange availability, so the period was one when restrictions were on the increase. At the end of June 1968, the import program for the second half of the year had, on net, transferred 35 items from the automatic-approval to the restricted list. Early in July, guarantee deposit requirements were extended to all imports, whereas previously they had been required only against letters of credit; the amounts of such deposit requirements ranged up to 200 percent. Later that month, regulations were changed for a variety of invisible transactions, and platinum trade, both import and export, was subjected to Ministry of Finance approval. Regulations for the trade of enterprises in export zones were also announced. In

TABLE 33 Number of Items in Each Import Regime,
1968–1975

	Prohibited	Restricted	Automatic Approval	Total
Second half 1967	118	402	792	1,312
Second half 1968	71	479	756	1,312
Second half 1969	74	530	708	1,312
Second half 1970	73	524	715	1,312
December 31, 1971	73	570	669	1,312
December 31, 1972	73	571	668	1,312
December 31, 1973	73	556	683	1,312
December 31, 1974	71	563	678	1,312
December 31, 1975	66	602	644	1,312

Sources: Frank, Kim, and Westphal, p. 59 for 1967 to 1970; IMF, *Annual Report on Exchange Restrictions,* various issues from 1971 to 1975.

Note: After 1967, the enumeration of items within lists was done on an SITC basis. The total, 1,312, represents the total number of SITC categories. These data were issued by the Ministry of Commerce and Industry. There are, of course, sub-categories within each major group. Thus, the 17,128 sub-items on the AA list after July 25, 1967, were from 792 items. See Frank, Kim and Westphal, pp. 58–59.

September, machinery imports through the government's Office of Supply were suspended, in the hope of encouraging substitution of domestic for foreign machines. Also, 134 sub-items of imports were added to the list which were subject to special tariffs of 70–90 percent at the discretion of the Ministry of Finance. Finally, in November, restrictions were imposed on importation of machinery from countries whose exports to Korea were more than double their imports.[15]

These moves tended to increase the restrictiveness of the regime to some extent. A few changes went the other way: some items which had been eligible for importation only from Japanese Property and Claims Funds were shifted to eligibility for importation from Korean foreign-exchange resources; exporters were permitted for the first time to engage in forward transactions in designated currencies, and foreigners entering Korea no longer had to register undesignated currencies and were permitted to exchange up to $100 upon departure without proof of original purchase.[16]

The pattern in other years was similar, both in that there were minor changes in many aspects of trade and payments and in the ways in which the restrictionist content of exchange control shifted with the current and prospective balance-of-payments situation.

In sum, the trade-and-payments regime from 1967 onward was much more liberalized than that which had prevailed in the 1950s or even the early 1960s. After that, however, further liberalization was sporadic and short-lived. In light of the very rapid growth of export earnings, and the government's announced program for continued liberalization, this failure to complete the task of removing quantitative restrictions is surprising. Frank, Kim, and Westphal provide a diagnosis as to the reasons for failure:

> Despite these and other attempts at further liberalization and reform, resort to the old price-distorting policies and controls was common. A number of factors were involved. First, any adverse trends in the balance of payments prompted a return to the old methods ... Secondly, as debt service payments began to rise, even though foreign exchange holdings seemed quite adequate in the late 1960's and early 1970's, concern over future debt repayments increased along with a fear for the vulnerability of the basic balance of payments. Restrictions on capital movements were strengthened in 1970. Finally, and probably most important, certain vested interests in the business community had much to lose from further liberalization and favored a return to price-distorting mechanisms. Since these interests wielded considerable political power, the tariff reform of 1967 wrought few real changes ... The business interests, many of them exporters who benefitted greatly from tariff exemptions and wastage allowances, exerted pressure through the Ministry of Commerce and Industry, and thus fostered a bureaucratic struggle between two ministries.[17]

EXPORT PERFORMANCE

Table 34 gives a breakdown of exports by sector of origin for the years 1966 to 1974. It should be remembered that many exports used a high proportion of imported intermediate inputs, so that the gross value of exports somewhat overstates their importance in this period. Nonetheless, it will be seen below that any reasonable adjustment for the growth of imported inputs still leaves an extremely high rate of growth of net exports.

As in the 1960–1965 period, rapid growth occurred in almost every sector. Only minerals exports were stagnant. The growth in Agriculture and Processed Food exports reflects primarily the rapid expansion of seafood exports. Exports originating in the fisheries and seafood-processing sectors were about $31 million in 1966 and rose to $166.1 million in 1973. The three industries that comprise the textile sector—Fiber Spinning, Textile Fabrics, and Textile Products—increased exports from about $80 million in 1966 to $1,431 million in 1974, thereby accounting for $1,350 million of the $4,208 million increment in total exports over the period. Thus, by 1974, despite very sizable growth rates in exports originating in virtually all sectors, textiles and their products accounted for 32 percent of all Korean exports.[18] Electrical Machinery, especially electronics, also expanded exports very rapidly, as did Plywood and the Miscellaneous Manufacturing sector. In all, there were exports of over $100 million in 1975 in each of the following categories: woven textile fabrics, electrical machinery· and appliances, miscellaneous manufactures, fish, plates and sheets of iron and steel, veneer sheets and plywood, footwear, transport equipment, manufactures of metal, and non-metallic mineral products.[19]

In dollar value, the compound annual rate of growth of exports over the 1966-to-1975 period was 40 percent. While part of that growth, especially after 1972, reflected worldwide inflation, the compound rate of growth of exports from 1966 to 1972 was 38 percent. The unit value index for Korea's exports, in dollar terms on a 1970 base, stood at 93.5 in 1966 and at

TABLE 34 Commodity Composition of Exports, 1966–1975 ($ millions)

	1966	1967	1968	1969	1970	1971	1972	1973	1974	1975
Agriculture, Forestry and Fisheries	30.7	32.2	38.6	49.0	57.2	62.8	78.9	126.1	182.9	n.a.
Minerals	26.6	28.8	33.8	35.5	43.4	36.7	33.7	42.8	61.0	n.a.
Processed Foods, Beverages and Tobacco	23.9	23.0	24.2	28.8	39.3	39.5	63.6	179.6	221.3	n.a.
Fiber Spinning	14.7	19.5	22.5	31.3	50.9	72.9	93.9	164.9	174.2	n.a.
Textile Fabrics	23.0	31.5	42.1	41.1	46.8	57.1	100.3	273.5	276.9	n.a.
Textile Products	41.9	73.5	125.7	176.9	232.5	337.5	466.0	798.2	976.9	n.a.
Lumber and Plywood	30.0	39.1	65.7	79.7	92.5	128.0	169.3	308.9	195.0	n.a.
Wood Products, including Paper, Printing and Publishing, and Furniture	2.0	3.0	4.0	4.7	8.3	8.8	29.6	95.0	97.0	n.a.
Leather Products	1.5	2.0	1.6	1.4	2.6	5.7	13.0	28.8	70.6	n.a.
Rubber Products	5.5	8.3	12.1	11.8	18.0	36.5	54.2	90.2	192.2	n.a.

TABLE 34 (continued)

	1966	1967	1968	1969	1970	1971	1972	1973	1974	1975
Chemicals	1.1	2.5	3.7	9.8	15.5	27.1	46.8	60.2	94.3	n.a.
Petroleum and Coal Products	0	0	.1	2.2	4.7	6.9	16.2	32.8	108.0	n.a.
Non-Metallic Minerals	1.6	1.0	.9	5.2	6.7	13.5	23.9	46.8	84.7	n.a.
Iron and Steel and Steel Products	7.8	1.8	1.1	4.8	13.1	24.2	91.9	182.1	436.5	n.a.
Metal Products	7.4	8.8	11.3	14.7	18.4	17.5	29.7	81.2	156.1	n.a.
Machinery	3.1	3.8	3.8	8.4	8.0	11.3	30.7	55.5	66.6	n.a.
Electrical Machinery	5.1	7.4	18.9	36.6	44.6	72.8	134.7	354.6	527.5	n.a.
Transport Equipment	1.2	3.2	1.7	8.0	9.5	7.5	15.8	28.1	129.4	n.a.
Miscellaneous Manufacturing	19.4	30.0	42.2	70.7	119.5	98.1	137.7	271.4	400.6	n.a.
TOTAL	247.6	320.3	455.2	622.6	835.2	1067.6	1632.6	3225.3	4456.2	5081

Source: Hong, *Factor Supply*, Table A-12.

Note: Total includes scrap and unclassifiables which are not listed separately.

100 in 1972. Part of that increase may have reflected improved quality of Korean exports and their increasing acceptance on world markets. However, even if the entire increase in export unit value from 1966 to 1972 reflected rising world prices and not quality improvement, the average annual increase in the volume of exports still exceeded 35 percent.

After 1972, export prices moved in line with world prices and inflation. The quantum and unit-value indexes (on a 1970 base) for Korean exports are shown in Table 35. It would appear that

TABLE 35 Quantum and Unit-Value Indexes
for Korean Exports, 1972–1975

	Quantum	Unit Value
1972	194.6	99.9
1973	305.2	126.5
1974	333.4	160.2
1975	410.0	148.4

only from 1973 to 1974 did the rate of growth of export volume diminish; indeed, from 1974 to 1975, growth of export earnings of 14 percent was achieved despite a decline in export prices of about 7 percent. Contrasted with the impact of world events on other developing countries, the Korean ability to maintain momentum and adapt to altered world economic conditions was truly remarkable.

By 1975, Korea was also diversifying her sales by geographic destination. In particular, considerable energy was devoted to the development of the export of "construction services." Under contracts, primarily with oil exporters of the Middle East, Koreans were undertaking to build roads, hospitals, apartment complexes, and other major construction projects. Under these contracts, Korean firms provided the management, supplied the Korean labor, and obtained most of the steel, cement and other construction materials from Korean factories. By the end of

1975, $1.8 billion in contracts had been signed, and there was great optimism about future prospects. To the extent that earnings from these service-sector exports were already realized in 1975, the rate of growth of commodity exports understates that of goods and services.

The rapid increase in exports is reflected in the national income accounts: exports of goods and services, which it will be recalled had been less than 2 percent of GNP in the 1950s, constituted 10.3 percent of GNP in 1966, 14.7 percent of GNP in 1970, 32.0 percent of GNP in 1973, and 30.2 percent in 1975. As mentioned earlier, however, part of that increase was spurious, as most exporters used large quantities of imported intermediate goods in their production processes. The "wastage allowance" incentives granted to exporters, combined with the fact that the exchange rate for imports was overvalued for extended periods, probably increased the relative attractiveness of imported intermediate goods compared to what would have been most efficient.

Because the wastage allowance provisions overstated the quantity of imported inputs required to produce exports, reliable figures on the import requirements for exports are not available. There are some data, published by the Ministry of Finance, which estimate imports used *directly* in the production of exports. These figures probably provide an overestimate of the direct input requirements of imports for exports. However, they neglect indirect import requirements for exporting. Whether these two offsetting errors result in an over- or underestimate of total import requirements is difficult to judge. They do, however, give some idea of the order of magnitude of import requirements.

Table 36 gives data on the value of imports used *directly* in export production, as well as the direct import-export ratio and the implied value of net exports for the 1963–1975 period. As can be seen, it is estimated that imported inputs for export did not become a significant factor until 1964. Over the next several years, imports for export mushroomed from about $7 million

TABLE 36 Estimation of Net Exports, 1963–1975

	Gross Value of Exports		Value of Imports for Export Production	Net Exports	Ratio of		
	Manufactures	Total			Net Exports to Gross		Net Exports to GNP
					Mfrs.	Total	
	($ millions)						
1963	58.9	86.8	0	86.8	1.00	1.00	4.9
1964	81.7	118.9	6.9	111.9	.92	.94	4.4
1965	130.5	175.0	10.4	164.6	.92	.94	6.1
1966	189.4	247.6	100.1	147.5	.47	.60	4.2
1967	258.4	320.3	134.5	185.8	.48	.58	4.6
1968	381.5	455.2	212.4	242.9	.45	.53	5.0
1969	536.2	622.6	297.2	325.3	.45	.52	5.5
1970	731.0	835.2	386.3	301.1	.47	.36	6.0
1971	964.8	1067.6	506.3	561.3	.48	.53	7.2
1972	1517.3	1632.6	608.0	1024.6	.60	.63	11.1
1973	3030.7	3225.3	1620.5	1604.8	.47	.50	15.3
1974	4207.7	4456.2	2111.9	2344.3	.50	.53	16.0
1975	4648.1	5081.0	2218.4	2862.6	.53	.56	15.6

Source: Wontack Hong, Statistical Appendix, Tables A-20 and A-21. The Ratio of Net Exports to GNP was derived by multiplying the ratio of gross exports to GNP, as given in the national income accounts, by the ratio of net to gross exports.

in 1964 to $100 million in 1966. Despite the fact that the government took measures to try to induce exporters to increase their utilization of domestically produced inputs,[20] there seems to be little doubt that imports for export rose even more rapidly than exports, at least until 1970. If one assumes that all imported intermediate goods were used in the production of manufactured exports, that would imply that the (direct) import content of manufactured exports was very close to 50 percent after 1966. Thus, the estimated growth *rate* of exports over the 1966–1972 period would not be affected by using net, rather than gross, export figures, and the estimated rate of growth from 1972 to 1975 would be reduced only slightly. Of course, to the extent that the data in Table 36 understate import requirements by neglecting the indirect component, this conclusion might be in error. However, when it is recalled that the wastage allowance probably provides an offsetting error in the other direction, it seems evident that correction for the large size of imported intermediate inputs for export does not significantly alter the conclusion that South Korea's export performance was truly remarkable.[21]

The 1966–1975 period witnessed changes in the geographic destination of Korea's exports, but many of these were reversed with the decade. The Japanese share, which had been falling sharply during the early 1960s, continued to decline until 1969 when it reached a low of 21.4 percent of Korean exports. Thereafter, it rose somewhat, but by no means reattained the level of the late 1950s. The U.S. share continued to rise in the late 1960s, reaching just over half of Korea's exports in 1968 and 1969, and then declined in the early 1970s. By 1975, the U.S. share of Korea's exports was 30.2 percent, compared to 35.2 percent in 1965 and 47.3 percent in 1970. Exports to Asian countries other than Japan fell from 23.9 percent of total exports in 1965 to 9.8 percent in 1970, but thereafter rose to 14.9 percent in 1975.[22] This was mostly offset by changes in the share of the rest of the world (not including Europe), notably the Middle East, whose share of Korean exports rose from 3.6 percent in 1965 to 10.9 percent in 1975.

IMPORTS

The combined use of both price and quantitative restrictions to influence both the total amount and the commodity composition of imports has already been discussed. The fluctuations in the real price of imports, as seen in the import PLD EERs in Table 32, were offset in part by variable tariff levies, altered guarantee deposits, and other measures.

Exporters' demands for imported intermediate inputs, capital goods import requirements, the price and quantitative measures discussed above, and the extent of inflationary pressure emanating from governmental monetary and fiscal policy all combined to influence the level of imports and their commodity composition. The data are given in Table 37.

As comparison with Table 28 reveals, the major shift in import composition from earlier years was an increased share of Machinery and Transport Equipment in total imports and an offsetting reduction in the share of Chemicals. To be sure, all categories of imports, at least at the level of aggregation presented here, grew rapidly in absolute amount. Even imports of Chemicals, whose share fell from over 20 to less than 10 percent of the total, rose from $103 million in 1965 to over $200 million in 1971 and 1972 and to $790 million in 1975.

It is not possible to associate the behavior of imports in various commodity categories with their classification in the import programs. In general, license applications for raw materials, intermediate goods, and capital goods were subject to automatic approval unless domestic productive capacity was extensive, in which case they were on the restricted list. Consumer goods deemed "essential" and not domestically produced—mostly food, beverages, and some manufactures—were also placed on the automatic approval list. The restricted list contained items

TABLE 37 Commodity Composition of Imports, 1966–1975
($ millions—% distribution in parentheses)

	Food and Beverages	Crude Materials	Mineral Fuels	Chemicals	Manufactures	Machinery and Transport Equipment	Total
1966	73 (10)	159 (22)	42 (6)	135 (19)	136 (19)	172 (24)	716
1967	95 (10)	215 (22)	62 (6)	113 (11)	201 (20)	310 (31)	996
1968	169 (12)	275 (19)	75 (5)	128 (9)	281 (19)	533 (36)	1463
1969	303 (17)	345 (19)	111 (6)	137 (7)	335 (18)	593 (33)	1824
1970	321 (16)	420 (21)	136 (7)	164 (8)	353 (18)	589 (30)	1984
1971	403 (17)	484 (20)	139 (8)	201 (8)	430 (18)	685 (29)	2394
1972	365 (14)	475 (19)	219 (9)	223 (9)	477 (19)	762 (30)	2522
1973	576 (14)	948 (22)	312 (7)	345 (8)	902 (21)	1157 (27)	4240
1974	829 (12)	1307 (19)	1054 (15)	631 (9)	1167 (17)	1848 (27)	6852
1975	959 (13)	1171 (16)	1387 (19)	790 (11)	1053 (14)	1909 (26)	7274

Source: BOK, Economic Statistics Yearbook, 1976.

Note: Figures do not add to totals due both to rounding and to the exclusion of a "not elsewhere classified category," which was as small as U.S. $30,000 in 1966 and as large as $14.2 million in 1974.

whose importation might compete with domestic production or whose utilization might cover a wide range of applications. Prohibited items, by and large, were "luxury goods" and goods deemed injurious to welfare.[23]

It should be noted that the rapid growth of imports resulted in an increased share of imports in GNP over the years of rapid export growth. It will be recalled that imports averaged just over 10 percent of GNP in the late 1950s. Their share rose somewhat in the early 1960s, reaching almost 16 percent in 1965. By 1968, imports of goods and services constituted 26 percent of GNP at current market prices, and they remained at about that level from then until 1972.[24] They then rose in relative importance still further, representing 43 percent of GNP in 1974 and 40 percent in 1975. Thus, the years of rapid export growth were accompanied by increased relative importance of imports. Part of this, as already mentioned, was the natural consequence of the high import content of exports. Much, however, reflected the increasing openness of the Korean economy.

One final aspect of the import regime deserves mention—the tariff structure and attempts to alter it. One of the remarkable aspects of Korea's trade-and-payments regime over the period since 1953 has been the remarkable stability of the tariff structure. It was seen in Chapter 2 that an attempt at tariff reform in 1957 left the basic structure fundamentally unchanged. With the devaluations of 1961 and 1964, "special tariffs" were imposed to absorb any premiums generated by the import programs on import licenses. Other than that, and changes in a very small number of tariff rates for specific commodities, the tariff regime remained basically unaltered until 1967.[25]

In that year, another effort at fundamental reform was undertaken. That reform was part of the attempt at that time to liberalize the regime, and the original intent of the reform appears to have been to lower the tariff rates on a wide variety of commodities. Table 38 gives the legal tariff rates for some categories derived from the old and the revised tariff schedules. As can be seen, the rate structure was little altered by the

TABLE 38 Representative Legal Tariff Rates
Before and After Tariff Reform, 1967
(simple averages of rates within sectors)

BTN Section	Old Rate %	New Rate %
4. Prepared foodstuffs, etc.	84.3	95.1
5. Mineral products	15.9	25.2
6. Products of chemical and allied industries	27.6	29.7
9. Wood and wood articles	40.1	44.2
11. Textiles and textile articles	59.0	71.0
15. Base metals and articles thereof	32.9	35.6
16. Machinery and mechanical appliances	27.4	30.6
20. Miscellaneous manufactured articles	78.9	81.9

Source: Frank, Kim, and Westphal, p. 60.

revision and, if anything, rates tended to rise rather than fall. The tariff reform had begun in response to a consultant's recommendation that the tariff structure be simplified, with a basic rate of about 20 percent on most imports and a special rate for a few commodities where there were special reasons for extra protection.[26] For the reasons discussed above, that aspect of the liberalization failed, as the groups benefiting from protection were able to exert enough political influence to ward off the reductions in tariff rates that would otherwise have been made.

Yet another change in the tariff structure was effected in February 1973. Some tariffs were raised and some lowered; the number of items subject to tariff increased from 3,174 to 3,985 while the average rate of duty fell from 38.8 to 31.3 percent.[27] The ratio of actual tariff collections to imports over the period does show a declining trend, evident in Table 39.[28]

Since actual collections originate only from imports subject to duty, it is difficult to estimate the degree to which the declining ratio reflects a lower average tariff rate or a higher fraction of duty-exempt imports. Moreover, these data do not fully reflect all charges against imports. There were "special tariffs" applicable

TABLE 39 Actual Tariff Collection and Imports, Selected Years

	1966	1968	1970	1971	1972	1973	1974
Imports ($ millions)	716.5	1464.1	1985.0	2395.0	2522.0	4241.5	6844.6
Tariffs collected ($ millions)	69.4	140.2	183.4	154.5	138.8	184.3	288.5
Tariffs/Imports	0.10	0.10	0.09	0.06	0.06	0.04	0.04

Source: See Foonote 28.

at rates of 70–80 percent for a large variety of goods on the restricted import lists, as discussed in Chapter 3. These rates were designed to absorb scarcity premiums on import licenses. In addition, the government had authority to alter tariff rates by 50 percent of c.i.f. value by administrative fiat.[29]

Despite all these qualifications, one conclusion emerges clearly: the actual duty rates collected were not very high. If one takes total duty collections as a percent of total imports in 1966, for example, they were 8 percent; even in 1968, the figure was only 10 percent. While these figures undoubtedly represent a weighted average of rates for imports subject to duties and for duty-exempt commodities, tariff rates were nonetheless relatively moderate throughout the 1966 to 1975 period.[30]

CAPITAL FLOWS

It was seen in Chapter 3 that policies to attract foreign capital inflows were implemented starting in the early 1960s. Prior to that time, there had been virtually no foreign direct investment or lending—commercial or public—to Korea.[31] In the years following the passage of the Foreign Capital Inducement Law, foreign capital flows increased markedly, although from a very small base. In 1966, a number of revisions were made in the law, designed to increase further the attractiveness of lending and investing in Korea. Changes made at that time included: the removal of any minimum requirement for Korean participation in equity capital; provision for governmental assumption of management responsibilities in the event that any foreign-financed firms threatened default; limitation of governmental guarantees so that debt service liabilities from them could not exceed 9 percent of annual foreign-exchange receipts (thereby insuring the worth of the governmental guarantee); and increased tax exemptions and tax holidays for foreign firms and investors.

While the Foreign Capital Inducement Law, as amended, increased the attractiveness of lending and investing in Korea, the interest rate reforms of 1965 were even more significant in attracting commercial lending to Korea and in making foreign borrowing attractive to Koreans.[32] The interest rate reforms had increased the rate of interest to about 26 percent for borrowing from domestic sources, with loans for favored prospects extended at 18 percent. However, the prevailing interest rate on dollar-denominated loans was about 12 percent, and those with government guarantees were extended at even lower rates. Foreign lenders were willing to lend large amounts to Korean firms, backed as they were by governmental guarantees and the rapid growth of the Korean economy and of foreign-exchange earnings.

The fact that the exchange rate applicable to capital-account transactions for the Korean wŏn remained relatively stationary over extended periods in the late 1960s meant that the real rate of interest paid by Korean firms for foreign loans was generally negative, and certainly far less than the rate paid on domestic borrowing. Indeed, it would have required an annual rate of depreciation of the currency of about 14 percent to equalize the attractiveness of domestic and foreign borrowing.[33] In fact, the average annual rate of currency depreciation was about 3.2 percent over the 1965–1970 period, and the actual nominal interest rate on foreign loans was between 5.6 and 7.1 percent.[34]

In these circumstances, it is not surprising that foreign borrowing mounted rapidly in the late 1960s. Table 40 gives data on new borrowing, interest and principal repayments, net borrowing, and net indebtedness, for the period from 1959 to 1975. As can be seen, there had been no borrowing before 1959. From 1959 to 1962, all borrowing was done by the government—primarily from the Development Loan Fund. As of the end of 1965, total indebtedness was $301 million, of which $176 million was public and the remainder private. Thereafter, borrowing and net indebtedness grew rapidly. Total indebtedness tripled between the end of 1965 and the end of 1967 and, by the end of 1971, was ten times the level it had been at the end of 1965. After 1965, borrowing by the government and private sectors was about equal in magnitude: by 1975, net government debt was $3,125 million and private debt was $3,571 million. Almost all of this was long term: at the end of 1975, the private sector owed only $284 million with a maturity of three years or less.

TABLE 40   Loans, Debt Service, and Net Indebtedness, Commitment Basis 1959–1975 ($ millions)

	New Borrowing	Debt Service and Repayment			Net Borrowing	Net Indebtedness
		Principal	Interest	Total		
1959	6.7	.0	.0	.0	6.7	6.7
1960	5.0	.2	.1	.3	4.8	11.6
1961	3.1	.4	.1	.5	2.6	14.3
1962	55.5	.6	.2	.8	54.7	69.2
1963	92.2	5.3	.6	5.9	86.3	156.0
1964	47.1	5.7	1.6	7.3	39.8	197.4
1965	111.1	7.2	2.7	9.9	101.2	301.3
1966	233.8	10.8	4.8	15.6	218.2	524.3
1967	455.6	25.6	10.2	35.8	419.8	954.3
1968	616.8	67.7	15.9	83.6	533.2	1568.4
1969	637.1	108.2	29.5	137.7	499.4	2097.4
1970	681.9	209.3	52.6	261.9	420.0	2570.0
1971	722.4	247.2	79.9	327.1	395.3	3044.2
1972	858.7	301.0	113.3	414.3	444.4	3601.9
1973	1224.4	344.4	160.9	505.3	719.1	4481.9
1974	1778.8	391.2	217.0	608.2	1170.6	5869.4
1975	1202.5	376.1	271.4	647.5	555.0	6695.8

Source: Data kindly provided by the Economic Planning Board.

By contrast, direct investment was relatively much smaller early in the decade and began increasing rapidly only in 1972. It rose from about $20 million in 1965 to $61 million in 1970 and $110 million in 1972, as can be seen in Table 41. Even as of June 1973, cumulative foreign direct investment approvals from

1966 were only $513 million, of which $301 million had arrived.
Of the $513 million approved, $305 million were from Japan
and $170 million from the United States. In value terms, the
sectors in which there was the most investment were textiles
and apparel ($126 million, 61 projects), electric and electronics
($84 million, 127 projects), and hotels and tourism ($52 million, 9 projects).[35] Thus, it was lending, and not direct investment, that provided a substantial source of foreign exchange in
the late 1960s. As foreign grant aid declined in both relative and
absolute importance, direct investment, commercial lending to
the private sector, and loans to the government partially replaced
it. Some of the government borrowing, of course, represented
concessional loans, discussed further below.

The value of exports is given in the fifth column of Table 41,
and the sixth and seventh give the debt-service ratio—principal
and interest repayments as a fraction of exports—and the ratio
of capital flows to exports for the 1965 to 1975 period. The
fact that South Korea had virtually no outstanding debt as of
1965 is reflected in her unusually low debt-service ratio of
0.057. The rapid increase in debt, and accompanying debt-
service obligations, resulted in a sharp increase in that ratio in
the late 1960s despite the rapid growth of exports. By 1971,
debt-servicing and repayments obligations stood at over 30
percent of export earnings—a high ratio by any standard.

This led some observers to question the soundness of Korea's
expansion during that period. In fact, the balance-of-payments
concerns in the late 1960s and early 1970s were attributable to
mounting questions about the debt. The real difficulty arose
because of the differential interest rate payable on foreign loans.
That, combined with the fixing of the exchange rate for
relatively long intervals, enabled speculation against currency
changes. It was in Korean businessmen's interest, if they could,
to borrow immediately after a devaluation but to attempt to
buy dollars to repay their loans prior to any anticipated devaluation. This phenomenon occurred, for example, prior to the
adjustment of the exchange rate to 370 wŏn per dollar in June

TABLE 41 Net Borrowing, Direct Investment, Debt Service,
And Export Earnings, 1965–1975
($ millions)

	Net Borrowing	Direct Investment	Total (1) + (2)	Debt Service	Exports	Debt Service Ratio (4)/(5)	Capital Flows/ Exports (3)/(5)
	(1)	(2)	(3)	(4)	(5)		
1965	101.2	20.1	121.3	9.9	175.1	.057	.69
1966	218.2	2.2	220.4	15.6	250.3	.062	.88
1967	419.8	19.9	439.7	35.8	320.2	.112	1.37
1968	533.2	24.2	557.4	83.6	455.4	.184	1.22
1969	499.2	28.2	527.4	137.7	622.5	.221	.85
1970	420.0	61.4	481.4	261.9	835.2	.314	.58
1971	395.3	45.2	440.5	327.1	1067.6	.306	.41
1972	444.4	110.4	554.8	414.3	1624.1	.255	.34
1973	719.1	264.7	983.8	505.3	3225.0	.157	.31
1974	1170.6	139.9	1310.5	608.2	4460.4	.136	.29
1975	555.0	n.a.	n.a.	647.5	5081.0	.127	n.a.

Sources: Same as Table 40 for columns (1) and (4). Exports from BOK, *Economic Statistics Yearbook* and Direct Investment from Wontack Hong, "Trade, Distortions, and Employment Growth," Table 4.15.

Note: Direct investment is recorded on a commitment basis. For the first ten months of 1975, direct investment approvals were $179 million. See Suk Tai Suh, "Statistical Report on Foreign Assistance and Loans to Korea," KDI Monograph 7602, (Mimeo, 1976), p. 71.

1971, as foreign-exchange reserves (which had been growing rapidly in earlier years) declined from a peak of $618 million in October 1970 to $573 million in May 1971. As inspection of Table 40 shows, the repayments of loans were sizable at about that time, and new borrowing had leveled off, thus resulting in a decline in net borrowing in 1970 and 1971.

As the last column in Table 41 indicates, rapid growth of exports prevented the emergence of what could otherwise have been a serious problem. Borrowing from abroad reached its peak—in relative importance—in 1967 and 1968. It was a very important offset to the drop in aid taking place at that time, as net foreign loans and investments were equal to 1.37 and 1.22 times export earnings in those two years. Thereafter, the rapid growth of export earnings outpaced the growth of indebtedness, and the relative importance of capital inflows as a source of foreign exchange dropped sharply. By 1972, foreign borrowing and investment were equal to only about one-third of export earnings. As mentioned above, it was at about that time that the authorities were struggling with debt-management problems and attempting to restructure their obligations.[36]

Public lending—much of which was at concessional terms— will be discussed later in this chapter. Here, focus is on commercial transactions. Of total private debt of $1,871 million in 1970, $111 million was short-term and the rest had maturities in excess of three years. A total of $488 million was held by American lenders, and $288 million by Japanese. The remainder—including the entire short-term debt—had originated from international financial markets. Thus, the relative importance of the United States and Japan was far less in Korea's access to international capital markets than in terms of her markets for her exports.[37]

Commercial loans were extended to all sectors of the economy, not simply manufacturing. Table 42 gives a breakdown of the sectoral destination of both commercial and public loans. As can be seen, Electricity, Transportation, and other social overhead capital sectors received almost as much from commercial sources as did Manufacturing. Within Manufacturing, Textiles, Chemicals, and Petroleum sectors were the largest commercial borrowers, accounting for 57 percent of manufacturing commercial borrowing. Since access to international capital markets was regulated by the government, these data reflect government-set priorities.

By the early 1970s, Korean credit-worthiness was recognized in international capital markets. Just as Korean exporters were establishing themselves in international commodity markets, Korean firms were developing their contacts and learning the vagaries of the international capital market.

This experience was of great importance in easing the adjustment to the shocks from the international economy in 1973–1974. When the oil price increased late in 1973, the potential harm to the Korean economy was massive: imports of petroleum and petroleum products had been $218 million in 1972 and were $296 million in 1973; they then rose to $1,020 million in 1974 —the increase alone represented an increase of 17 percent in the 1973 level of total imports. Moreover, unlike some countries where there were offsetting and important increases in export prices to cushion the impact, there was little such effect for Korea: import prices (including raw materials in addition to oil) rose 107 percent from 1972 to 1974, while export prices increased only 60 percent, much of which must have reflected the increased price of imported inputs used in exports.

If South Korea had exported and imported the same quantities in 1974 as in 1972, her export earnings would have been 1,305 billion wŏn and her 1972 imports would have cost 2,104 billion wŏn—an increment of 600 billion wŏn over the actual trade deficit in that year, compared with a GNP of 3,875 billion wŏn. This represented 15.5 percent of 1972 GNP—a huge amount. Had the adjustment had to be completed immediately, it would have been extremely difficult. In fact, Korea was able to increase her net borrowing from abroad by over $275 million in 1973 and by another $450 million in 1974 (see Table 41),

TABLE 42 Sectoral Distribution of Foreign Loans, 1959–1975 ($1,000s)

	Commercial Loans				Public Loans			
	1959–1966	1967–1971	1972–1975	TOTAL	1959–1966	1967–1971	1972–1975	TOTAL
Primary Sector (1–6)	52.3	195.2	194.8	442.3	14.1	71.0	242.8	327.9
Manufacturing Sector	220.2	2368.5	5106.9	7695.7	39.7	272.3	663.7	975.6
Textiles (10–12)	65.2	462.1	722.9	1250.2	13.8	31.2	23.2	68.3
Textile Fibres (10)	40.0	389.9	573.9	1003.8	13.8	31.2	23.2	68.3
Lumber and Plywood (13)	0.4	47.0	39.5	86.8	—[a]	—	—	—
Paper and Products (15)	1.2	50.1	31.7	83.5	—	—	—	—
Chemicals (19–21)	43.6	499.3	1228.0	1770.9	13.9	186.4	300.9	501.2
Basics & Fertilizers (19 + 21)	40.0	257.7	476.1	773.9	13.9	186.4	300.9	501.2
Petroleum (22)	38.8	491.1	846.8	1376.6	—	—	—	—
Metals (25–26, 28)	3.8	297.2	798.8	1009.8	—	8.3	253.2	261.5
Iron & Steel (25)	—	80.8	337.0	417.8	—	8.3	253.2	261.5
Non-Metallic (24, 27)	51.1	408.7	520.8	980.6	12.0	34.8	27.3	74.0
Non-Metallic Minerals (24)	51.1	365.2	474.6	890.9	12.0	34.8	27.3	74.0
Machinery (29, 30)	12.5	55.2	100.4	168.1	—	8.2	47.4	55.6
Non-Electrical (29)	0.04	10.7	5.2	16.0	—	8.2	47.4	55.6
Transport Equipment (31)	2.9	84.6	652.8	740.3	—	—	—	—

TABLE 42 (continued)

	1959–1966	1967–1971	1972–1975	TOTAL	1959–1966	1967–1971	1972–1975	TOTAL
	Commercial Loans				Public Loans			
Other Mfg. (32 + the rest of 7–32)	0.3	63.2	165.3	228.8	—a	3.4	11.6	14.9
Social Overhead Capital and Service	12.7	1367.3	2462.0	3842.0	183.8	1055.1	2779.7	4018.5
Electricity (35)	2.1	694.1	1356.5	2052.7	26.3	277.6	523.0	826.9
Transportation (39)	3.0	542.7	942.3	1488.1	104.6	525.5	1424.7	2054.8
Others	7.6	130.5	163.1	301.2	52.8	252.0	832.0	1136.8
All Sectors	285.2	3931.0	8025.8	12242.0	254.6	2377.9	7785.5	10417.9

Source: Wontack Hong, Statistical Data, Tables B-29 and B-32.

Note:a— means data were not separately listed for the sector. Amounts in those sectors were presumably small.

thus enabling a much smoother adjustment than would otherwise have been possible. To be sure, there were moments of doubt as to whether all debt-service obligations could be met and moments when foreign lenders' confidence appeared somewhat shaken. Nonetheless, Korea's ability to borrow abroad was a necessary condition for her to weather the oil price increase and resume growth as rapidly as she in fact did. The ability to borrow abroad gave policy-makers the time to alter instruments of domestic policy and to adapt to altered circumstances. Had Korea instead been forced to curtail imports abruptly, the resulting dislocations would have prevented the rapid resumption of export growth. The fact that Korea was credit-worthy, was not, of course, fortuitous. On the contrary, it was her earlier borrowing-and-repayment history, itself the result of the export-promotion strategy, that stood her in such good stead.

AID AND OFFICIAL CAPITAL FLOWS

It was seen in Chapter 3 that aid was already diminishing in both absolute and relative importance in the early 1960s. That trend continued during the latter half of the 1960s. In addition, the United States lost its position as the sole provider of aid and switched most of its aid from grants to loans. After 1972, grant aid ceased, and total U.S. assistance—PL 480 and loans—dropped sharply.

Table 43 gives the overall outlines. Comparison of those data with the numbers in Table 31 indicates that total American aid, including loans, grants, and PL 480 sales, was fairly steady from 1966 to 1972, and then declined sharply. The totals, however, do not tell the full story. Non-project supporting assistance, which had been the largest single component of aid in the late 1950s, fell continuously until, by 1972, it was negligible. Even project assistance, which had not been nearly as important as supporting assistance, fell to levels of around $5 million annually in the late 1960s and phased out in the early 1970s. Until the

TABLE 43 Total Aid Flows From the United States,
1966–1975
($ millions)

	1966	1967	1968	1969	1970	1971	1972	1973	1974	1975
Non-Project Supporting Assistance	54.8	59.8	43.7	16.7	14.2	9.4	.6	—	—	—
Project Assistance	5.2	5.6	9.9	7.5	6.4	5.1	3.4	3.3	2.1	.9
PL 480 Sales	35.0	58.0	58.5	64.9	54.7	30.8	3.7	—	—	—
PL 480 Loans	28.5	31.6	47.3	118.5	67.8	84.9	197.0	61.0	—	84.0
Development Loans	49.7	74.8	38.0	31.9	38.8	55.7	36.8	27.7	44.1	192.2
TOTAL	173.2	229.8	197.5	239.7	181.8	185.9	241.4	92.1	46.2	277.0

Source: USAID to Korea. See note to Table 31.

early 1970s, PL 480 sales remained at about their levels of the early 1960s, and the two loan categories—PL 480 and development loans—rose sharply, thus offsetting the decline in other aid categories.

It is difficult to obtain comparable data on aid from other countries and international institutions because the Korean government treats all loans as being in the same category, regardless of whether terms are concessional or not. The one sizable official transfer that might be regarded—in terms of its economic impact—as having the same effect as aid was the Japanese settlement of 1965. Under the terms of that agreement, the Japanese government was to provide $300 million in grants and $200 million in public loans over the ten-year period from 1965 to 1975.[38] This, from the Korean viewpoint, represented a reparation settlement and did not constitute aid. It nonetheless provided a sizable transfer of resources which supplemented those from foreign borrowing, foreign investment, and current account earnings.[39]

There are two ways in which one can attempt to estimate the volume of concessional loans received during the late 1960s and early 1970s. One is to regard all government borrowing as concessional. A breakdown by source of government loans is also available so one can also get an idea of the relative importance of the United States and Japan. The loan data are shown in Table 44. These numbers would suggest that, at least before 1969, concessional loans from sources other than the United States and Japan could not have been a significant factor in total foreign exchange availability, since the residual is fairly small.

The second piece of evidence consists of data on the interest-rate structure and grace periods for public and private loans. Those data are given in Table 45 and cover all borrowing from 1959 to 1974, without a breakdown by year. Comparison of the structure of interest rates, the grace period, and the repayment period strongly suggests that probably at least three-quarters of all public borrowing was concessional: less than one-third was subject to interest rates in excess of 5 percent, and the weighted

TABLE 44 Government Borrowing, 1966–1975
($ millions)

	1966	1967	1968	1969	1970	1971	1972	1973	1974	1975
Net Government										
Borrowing	127.5	105.0	102.4	155.9	288.2	323.2	565.9	596.6	446.9	586.9
Japanese Loans	44.3	27.2	17.9	11.1	88.5	26.3	176.7	106.4	175.7	0.7
American Loans	64.4	64.0	79.5	61.4	55.1	124.4	275.4	188.4	35.0	102.8
Residual	18.8	13.8	5.0	83.4	144.6	172.5	113.8	301.8	236.2	483.4

Source: Same as Table 40.

TABLE 45 Foreign Loans by Interest Rates and Terms of
Repayment, 1959–1974
(commitment basis)

	Commercial Loans		Public Loans	
	$ Million	*% Share*	*$ Million*	*% Share*
Interest Rate (Weighted Av. = 7.1%)			(Weighted Av. = 4.1%)	
0–1%			173.6	(6.3%)
1–3%	31.5	(0.8%)	972.0	(35.2%)
3–4%			711.2	(25.7%)
4–5%	97.6	(2.4%)		
5–6%	1,418.6	(34.0%)	134.0	(4.8%)
6–7%	791.7	(19.0%)		
7–8%	426.5	(10.2%)		
8–9%	200.9	(4.8%)	767.4	(27.8%)
over 9%	182.9	(4.4%)		
floating rate	1,017.7	(24.4%)	6.2	(0.2%)
Grace Period (Weighted Av. = 2.5 Yrs.)			(Weighted Av. = 7.2 Yrs.)	
0–1 Years	793.5	(19.0%)		
1–2 Years	1,008.1	(24.2%)		
2–3 Years	1,189.9	(28.6%)	782.3	(28.3%)
3–4 Years	628.4	(15.1%)		
4–5 Years	212.0	(5.1%)		
5–9 Years	334.4	(8.0%)	661.2	(23.9%)
over 9 Years	—	—	1,320.9	(47.8%)
Repayment Period				
(Weighted Av. = 10.1 Yrs.)			(Weighted Av. = 26.0 Yrs.)	
3–10 Years	1,957.5	(47.0%)	10.5	(0.4%)
10–15 Years	1,872.3	(44.9%)	937.5	(33.9%)
15–20 Years	336.7	(8.1%)		
20–30 Years			698.0	(25.3%)
30–40 Years			1,006.5	(36.4%)

TABLE 45 (continued)

| | Commercial Loans | | Public Loans | |
	$ Million	% Share	$ Million	% Share
Repayment Period (continued)				
over 40 Years			111.9	(4.0%)
Total	4,166.5	(100.0%)	2,764.4	(100.0%)

Source: EPB, "Foreign Capital Inducement and Investment Policy," (Mimeo, 1975).

average interest rate on loans to the government was 4.1 percent (compared with 7.1 percent for loans to the private sector). In confirmation of this, almost three-quarters of loans received by the government had a grace period in excess of five years, with very long repayment periods.

It would thus appear that, as a first approximation, most loans to the Korean government during the 1966–1972 period can be regarded as concessional.[40] That being the case, the data provided in Table 42 give some idea of the sectoral destination of those loans. The agricultural products loans are undoubtedly financed primarily from PL 480. It is apparent that a much higher fraction of public loans than of private loans was used to finance social overhead capital and that, contrary to the situation with regard to private loans, more than half went to support activities not clearly identified by sector. Thus, of the $845 million loans to the Korean government used to finance social overhead activities, less than $300 million were directed toward electricity, transport, and communication.

One other aspect of the change in aid patterns between the early 1960s and the later years should also be noted: as the data given above indicate, Americans and Japanese were the major sources of loans to the Korean government up until about 1968. Thereafter, their relative importance diminished as other lenders extended credit, primarily concessional, to the Korean government. Not only was the sum of aid—both loans and grants—smaller than had been the case in the late 1950s and the composition of aid different, with much greater proportion of it consisting of concessional loans, but the relative importance of the direct American contribution had decreased markedly. This was, of course, in accordance with American policy of shifting its aid toward the international lending agencies; but, even so, it marked the continuation of the trend toward reduced Korean interdependence with the United States.

Thus, as with other aspects of Korea's international economic relations, the period of the late 1960s and early 1970s saw the emergence of Korea with a much more balanced and diversified set of international links to nations than had earlier characterized her international economic relations. She was no longer crucially dependent on aid, and was able to enter the international financial markets to seek desired financing. When concessional loans were available, they cut the cost of borrowing but did not represent the only feasible source of foreign exchange. Concessional loans originated from diverse sources, not simply the United States. In marked contrast even to ten years before, it was no longer the case that the only available source of financing was concessional aid, and there was no longer heavy reliance upon one country or one source of financing.

Thus, the thirty-year modernization history witnessed, as one of its major aspects, Korea's transformation from an economically weak aid recipient, with links to the international economy primarily via her dependence on the United States and aid-financed imports, to an economically viable major exporting nation fully able to enter into international commercial and financial markets on commercial terms.

Obviously, elimination of the extreme dependence that prevailed in the 1950s was a prerequisite for modernization; the questions of how the trade-and-aid sector contributed to overall growth and of the efficiency of trade and aid are the topics of Chapter 5.

FIVE

The Allocative Efficiency of Trade and Aid

In an important sense, all four preceding chapters have been partially addressed to the topic that constitutes the focus of this chapter. It is impossible to chronicle the history of aid without, at least implicitly, providing some indication of its role in resource allocation. Any documentation of the growth of exports in the 1960s of necessity conveys a considerable amount of information about the efficiency of the export-promotion strategy.

It is not the purpose of this chapter to tread that ground again. Much of the analysis needed to evaluate microeconomic aspects of trade, aid, and capital flows has already been undertaken. Nonetheless, a variety of questions remain. To what extent was aid efficiently allocated and utilized? Could its timing or magnitude have been improved upon? Was the trade-and-payments regime optimal, or could variations in policy have enabled still better performance? While definitive quantitative answers to these and related questions are not possible, it is nonetheless worthwhile to provide some estimates and to assess the available evidence with respect to the efficiency of trade policy and aid.

This chapter focuses on the allocative efficiency of the trade regime, capital flows, and aid. The macroeconomic features, the contribution of foreign resources, exports, and other trade flows to aggregate growth, employment, and output levels will be examined in Chapter 6.

EFFICIENCY OF THE TRADE REGIME, 1945–1959

TRADE AND PAYMENTS, 1945–1959

There is little that needs to be said about trade and payments in the 1945–1950 period. Decisions were so completely dominated by short-term exigencies that it is difficult to determine what feasible alternatives might have been. Moreover, even if one could identify ways in which things might have been done better in the short run, there is the consideration that the benefits of such improvements would, in any event, have been lost during the Korean War.

As for the Korean War years, many of the same statements apply—in the conditions of those years, the concept of efficiency makes little sense. What *was* unfortunate during the war years was exchange-rate policy. While the physical controls on trade (enforced by military control of the ports) undoubtedly prevented the realization of desired transactions at the increasingly overvalued exchange rate, a pattern was set. And, if one were to pinpoint the single most important source of inefficiency during the 1954–1959 period, it was without question the effects—both direct and indirect—of the overvaluation of the exchange rate, and especially the response to incentives that were created. Those effects centered on the pattern of import substitution, the stagnation of exports, and the competition for the enormous profits accruing to those who did obtain imported goods; the resources devoted to those activities and the corruption that resulted diverted considerable resources from the reconstruction effort.

As seen above, exchange-rate policy resulted from the interaction of aid policies, the military effort, and the Korean government. It is the effects of those policies on the trade-and-payments regime, and the interaction of the regime and the import-substitution policies of the 1950s, that are of concern here.

Lack of data, especially reliable data, makes it extremely difficult to attempt any quantitative assessment of the magnitude of distortions caused by the "foreign exchange shortage" of the 1950s. Certainly there were large differentials between the landed cost of imports and the prices they could command in the domestic market; the tariff structure itself fails to convey an accurate impression of the magnitude of protection accorded domestic industry due to import restrictions, since not all who wished were able to obtain foreign exchange for imports at prevailing prices. For commodities not domestically produced, scarcity premiums for permitted imports were often sizable and constituted large prizes for those fortunate enough to get import rights.

There is no entirely satisfactory way of estimating the size or importance of these premiums. As seen in Chapter 2, an attempt to compare price series for comparable commodities from unit-value import statistics and domestic wholesale-price data was relatively unsuccessful. Lack of product homogeneity, erratic year-to-year fluctuations in unit value statistics, changes in the quality and/or specification of product for which a domestic price quotation was available, and other data problems were of sufficient magnitude to render these data of very questionable value. For cement, the estimated markup over landed cost appeared to have been in excess of 100 percent, but it was the only commodity for which data problems were not overwhelming.

It seems preferable, therefore, to eschew such data and to rely instead upon estimates of import volume relative to GNP for guidance as to orders of magnitude. Starting with these data, and estimating a "normal" price and share of imports in GNP, one can estimate the domestic price that would have been necessary to clear the domestic market as a function of the price elasticity of demand for imports. That price, in turn, can be compared with estimates of the real exchange rate for imports, and the differential can be taken as an estimate of the premium accruing to foreign exchange recipients.

In order to carry out the computation, the first problem was choice of a "normal" year; 1970 was chosen. It was assumed that import demand in that year, net of imports for re-export, constituted a "normal" percentage of GNP, and that the real exchange rate prevailing in that year was approximately at equilibrium. With those assumptions, it was possible to take data from the 1950s and to estimate the probable value of premiums on foreign exchange. The computations are given in Table 46. Row A gives imports as a percent of GNP, as estimated in the national income accounts, at constant 1970 wŏn. Row B gives exports as a percent of GNP, again in constant 1970 wŏn. Row C then provides the basic estimate of "domestic" imports, that is, those satisfying domestic demand.[1] These are defined as total imports less imports employed in producing exports. As can be seen, "domestic" imports in 1970 were 17.45 percent of GNP, compared with a range of 8 to 13.5 percent over the period 1954 to 1959.

The price of imports necessary to clear the domestic market with the actual volume of imports was then estimated on three alternative assumptions about the magnitude of the price elasticity of demand for imports. That is, it was assumed that, if imports had constituted 17.45 percent of GNP in any year in the 1950s, a domestic price of 260.1 (1970) wŏn per dollar in the domestic market would have equated the domestic price with landed cost. To be sure, this assumption is somewhat arbitrary, but it is not evident in which direction error may lie.

TABLE 46 Estimates of Scarcity Value of Imports, 1954–1959

	1954	1955	1956	1957	1958	1959	1970
A. Imports as % of GNP, (constant 1970 wŏn)	8.8	11.2	13.0	14.3	11.7	9.3	24.8
B. Exports as % of GNP, (constant 1970 wŏn)	1.1	1.4	1.2	1.5	1.9	2.1	14.7
C. "Domestic" imports (A - .5B) as % of GNP	8.15	10.50	12.40	13.55	10.75	8.25	17.45
D. Price of $1 of 1970 imports with demand elasticity of:							
D1. 1/2	1113.8	864.6	732.0	669.8	844.4	1100.0	260.1
D2. 1	556.9	432.3	366.0	334.9	422.2	550.1	260.1
D3. 2	408.5	346.2	313.1	292.5	341.1	405.1	260.1
E. Actual import PPP PLD EER	130.0	130.5	149.6	135.5	155.9	198.2	260.1
F. Scarcity premium (in 1970 wŏn) per dollar of imports							
F1. (D1 - E)	983.8	734.1	582.4	534.3	688.5	901.8	0
F2. (D2 - E)	426.9	301.8	216.4	199.4	266.3	351.9	0
F3. (D3 - E)	278.5	215.7	163.5	157.0	185.2	206.9	0

TABLE 46 (continued)

	1954	1955	1956	1957	1958	1959	1970
G. Scarcity premium as % of EER							
G1. (F1 ÷ E) × 100	757	562	389	394	441	454	0
G2. (F2 ÷ E) × 100	328	231	143	147	170	177	0
G3. (F3 ÷ E) × 100	214	165	109	116	119	104	0
H. Premiums as % of GNP							
H1. (0.01 × G1 × C)	61.7	59.0	48.2	53.4	47.5	37.5	0
H2. (0.01 × G2 × C)	26.7	24.3	17.7	19.9	18.4	14.6	0
H3. (0.01 × G3 × C)	17.4	17.3	13.5	15.7	12.8	8.6	0

On the one hand, it can be argued that the rising share of imports in GNP in the late 1960s may have reflected, at least in part, the increasing specialization of the Korean economy; on the other, it can be equally forcefully maintained that the low level of productive capacity in the Korean economy in the 1950s made the economy even more "dependent" upon imports than in the 1960s. To the extent the latter argument is valid, it might be that the share of imports in GNP in the 1950s would have had to be higher than in the 1960s for the same real exchange rate. For present purposes, the main point is that the estimates of premiums are biased upward if it is believed that the income elasticity of demand for imports in Korea is less than unity or that an earlier year should be chosen as indicative of "normalcy," and estimates of premiums are biased downward if it is believed that the income elasticity of demand for imports is greater than unity, or that a later year than 1970 should be used as a base.[2]

Rows D1, D2, and D3 provide estimates of the market-clearing price (in 1970 wŏn) of a dollar's worth of imports under the assumptions that the demand elasticity (for imports, not importables) was 1/2, 1, and 2.[3] Obviously, the lower the demand elasticity, the higher the domestic price to clear the market would have had to be.[4]

Row E gives the import PPP PLD EERs for 1954 to 1959. It will be recalled (from Chapter 2) that estimates of PPP PLD EERs for imports are not available for years prior to 1955. The number for 1954 represents the author's estimates of the highest the real import exchange rate could have been, given available information about the nature of the trade regime.

The three rows under F then express the scarcity premium per dollar of imports as the difference between the domestic price (as calculated by the manner described under D) and the landed cost (represented by the PPP PLD EER). For ease of comparison, those scarcity premiums are expressed as a fraction of the real EER in rows G, and rows H provide an estimate of the value of premiums (the ratio of imports to GNP times the premiums as a fraction of landed cost).

The most obvious characteristic of the estimates in Table 46 is the very large orders of magnitude implied by them. According to the estimates, there was no year when the scarcity premium on import licenses might have been less than the real exchange rate itself. The estimates conform with the impression from other sources—that the relative importance of scarcity premiums was declining throughout the 1950s. Even with the highest elasticity of demand, however, scarcity premiums are estimated to have constituted about 17 percent of GNP in 1954, falling to about 8.6 percent in 1959.[5]

While the numbers given in Table 46 are far too hypothetical to be taken as definitive, they provide still further basis for the belief that the scarcity of imports in the 1950s, and the value attaching to obtaining licenses, was of sufficient magnitude to have constituted a major distorting influence within the economy. With premiums of even half the estimated height, import licenses were surely very valuable. The economic costs associated with individuals' efforts to obtain those licenses were surely substantial. The resulting resource misallocation must be a factor of considerable importance in explaining Korea's economic performance in the 1950s.

Accounts of the 1950s by Koreans confirm that corruption was widespread and that political activity was centered upon obtaining rights to imports. This focus must have been detrimental to the reconstruction effort, since an ability to obtain imports at the prevailing EER must have been substantially more profitable than undertaking to expand production capacity. While data are unavailable, it is a reasonable conjecture that part of the rapid growth after 1960 may have resulted from the reallocation of resources previously allocated to trading in imports toward socially more profitable ventures.

EFFICIENCY OF IMPORT SUBSTITUTION
IN THE 1950s

The sizable premiums on imports accrued largely on items for which domestic production either had not started or was manifestly unable to meet more than a fraction of domestic demand. For, in the context of foreign exchange shortage, the authorities were reluctant to allocate foreign exchange toward items that could be domestically produced.

The consequence, of course, was relatively high levels of protection of those domestic industries that sprang up in response to the incentives provided by the trade-and-payments regime. To judge from accounts of contemporary observers, domestic industries in the import-substitution years displayed most of the characteristics of import-substitution industries in other countries: generally poor quality control, high levels of under-utilized capacity, heavy dependence on imports of critical inputs, use of relatively capital-intensive techniques, relatively slow growth of employment in manufacturing, and so on.[6]

An important question is the extent to which the import-substitution phase was a necessary or desirable precondition for later export-based growth. Unfortunately, data are not available on which to base a definitive judgment—no estimates of implicit levels of protection in the 1950s are available, and it is impossible to provide quantitative estimates of the extent to which the newly started import-substitution industries had high costs.

Several considerations are relevant, however, to putting some limits on the range within which the truth must lie. On the one hand, a number of factors served to keep the inefficiencies associated with import substitution within bounds and to render much of the investment economically justified. On the other hand, some pieces of evidence strongly suggest that there were excesses and inefficiencies that could have been avoided and that, therefore, there were aspects of the program that were wasteful and in no way contributed to later rapid growth.

To turn first to considerations suggesting that import substitution was probably not as excessive as it has been in some other instances, there is, first of all, the fact that it was less than a decade from the end of the Korean War, when the policy started, to the early 1960s, when it was abandoned. Starting from a very low base, the seven-year period of import substitution was simply not long enough to induce the development of high-cost intermediate and capital-goods industries. Indeed, given the destruction that accompanied the Korean War, a sizable fraction of total investment in the early years was directed to rebuilding destroyed industries that had formerly supplied the domestic market. It was seen in Chapter 2 that, despite the fact that manufacturing growth was largely accounted for by import substitution, the share of imports in total consumption by sector did not decline significantly. This was because much of the growth in output was destined to satisfy pent-up domestic demand.

It was probably these phenomena that accounted for the relatively painless transition to exporting in the early 1960s. If large investments had been made in high-cost capacity for intermediate and capital goods in the 1950s, substantial disruption would have resulted from enabling exporters to obtain raw materials and intermediate goods at world prices. Yet this access was necessary for the export drive of the 1960s: had the government been unwilling to witness the decline of domestic intermediate goods industries, their high-cost structure would have provided a substantial impediment to the export-promotion strategy. For purposes of understanding the way in which import substitution did or did not provide a foundation for later export-oriented growth, the most important consideration is probably the fact that import substitution did not, in any essential way, proceed to intermediate goods. It was therefore not difficult to open the domestic market to imports of those commodities to be used in export production, and there are no mentions, in contemporary accounts, of adjustment difficulties in domestic industries.

While all the factors listed above suggest that the excess costs of import substitution were probably far less than have been experienced in other circumstances, and when the policy has been carried out for a longer period of time, that does not prove that there were no excesses. Indeed, the very height of the estimated premiums on imports and knowledge of the magnitude of implicit protection provided for domestic industry indicate that there must have been substantial variance in domestic cost levels relative to international cost levels both between firms, within industries, and between industries. About the only indication that can be gleaned derives from production data. In general, after a change in incentives, one would expect that high-cost industries' outputs would contract for awhile, whereas industries able to compete successfully would continue expansion. Their expansion per se could then be taken as evidence that earlier investment had been economic.

One partial test of the extent of extreme inefficiency is, therefore, to examine the number of industries that did not reattain their production levels of the 1950s by a particular date.[7] For purposes of evaluation, the year 1964 was chosen. By that year, real GNP already exceeded its highest level of the 1950s by more than 25 percent. Industries whose output remained at levels lower than those attained in the 1950s can therefore be judged to have had significant difficulties in adjusting to altered incentives. That excess capacity remained provides a strong presumption that investment had been uneconomic and did not contribute to later development.

The *Economic Statistics Yearbook* provides production data for the 1955–1964 period for 101 manufacturing industries.[8] Of those industries, there were 20 for which the highest level of output reached in the 1950s exceeded that of the early 1960s, and had not been reattained by 1964. Those 20 must consist primarily of adversely affected high-cost import-substitution industries. Production of ethyl alcohol, for example, averaged about 30 thousand kiloliters in the late 1950s and fell to about half that level by 1964. Similarly, production of leather soles for shoes reached 5.7 million p'yŏng in 1958, and never exceeded 3.9 million between 1960 and 1964. Output of glass products peaked at 39,000 metric tons in 1956, and fell to 11,000 in 1962, reaching 24,000 by 1964. While some of the other declines in production were not proportionately as severe as these, many were. It must be concluded that a majority of these industries were probably ill-advised import-substitution ventures. To be sure, some may have resumed growth after 1974, but idle capacity over a five-year period is presumptive evidence of considerable inefficiency.[9]

Nonetheless, the fact that only about 20 percent of the industries covered experienced sustained declines is suggestive that excesses were limited. It is probable that other industries were able to adjust to changed conditions. Obviously, counting the number of industries whose output declined does not indicate their relative importance. In the absence of value statistics, a reasonable weighting system cannot be devised. Suffice it to say that, based on the data, this author would judge that the 20 declining industries' share of manufacturing value added was probably substantially below their unweighted proportion of industries. Certainly, for the industries whose output was still below levels of the 1950s by 1964, considerable investment must have been wasted during the import-substitution years. However, as a fraction of total investment and output in the 1950s, it would appear that the majority of industries were able to adapt to the altered incentives of the 1960s.

MICROECONOMIC ASPECTS OF AID

Two microeconomic aspects of aid must be separately assessed: the sectoral distribution of aid among economic activities; and the interaction of aid officials with the Korean government, which influenced policies affecting resource allocation.

It is always difficult to evaluate the sectoral allocation of aid. On the one hand, there is the consideration that, insofar as aid flows are directed toward specific activities, the recipient can redirect his own resources so that the net impact of aid in allocating resources may be quite different from that which an accounting of aid allocations would indicate. On the other hand, even if aid could not be interchanged with domestic resources, it is difficult to devise criteria with which to assess the optimality of the sectoral distribution of aid.

The situation in Korea in the 1940s and 1950s makes any microeconomic evaluation even more difficult. It was seen in Chapter 2 that aid really financed virtually all imports in the 1950s. Because it was so large, aid must be evaluated in terms of the efficiency of overall resource allocation in the 1950s. Even that allocation was affected primarily by overvaluation of the exchange rate and the accompanying import-substitution policies, so that it is primarily in terms of aid interaction with Korean government officials that the microeconomic aspects of aid must be discussed.

Before that is considered, however, a few salient features can be noted. First, aid emphasis upon education, both in the 1940s prior to the Korean War, and again during the reconstruction years, was crucial to building a foundation upon which the later success of the export-promotion policies could be based. Virtually all observers credit the quality of the Korean labor force as being a major permissive factor in enabling rapid growth; to a large extent, it was earlier aid efforts that provided such a basis. Second, there can be little doubt that relief supplies, both from 1945 to 1948 and from 1950 to 1954, were appropriately directed toward consumer goods and basic agricultural inputs; no other policy would have made sense. Finally, the role of the U.S. Military Government in undertaking land reform should not be discounted when it comes to evaluating the effects of aid upon Korean growth.

Thus, if fault can be found with some aspects of the aid program, and particularly the resulting exchange-rate policy, in the late 1950s, such a finding does not in any sense imply the failure of aid; indeed, in both its macro- and microeconomic aspects, aid was essential in buying time for Korea to reconstruct and develop.

It was in formulating economic policy that difficulties arose in the 1950s. As discussed earlier, the most important source of resource misallocation and microeconomic inefficiency lay in the complex and misguided exchange-rate policies of the 1950s. Had exchange-rate policy been more realistic, there would have been fewer irrationalities in import substitution than there in fact were.

It was seen in Chapter 2 that the pattern for currency over-valuation was set during the Korean War and had unfortunate consequences for the entire decade. It is fully understandable why and how, in the midst of wartime exigencies and in the absence of accumulated experience with the diplomacy of aid (from both the donor and the recipient viewpoint), the exchange-rate question became such a political football. Nonetheless, with full benefit of hindsight, it is obvious that settlement of military obligations in a way that did not provide the Korean government with a strong incentive to leave the currency overvalued would have been vastly superior to the practices actually followed. Once those practices started, currency overvaluation, with all its consequences, became a major problem for aid negotiations and economic policy during the rest of the 1950s.[10]

EFFICIENCY OF THE
EXPORT-PROMOTION STRATEGY

EXCHANGE-RATE POLICY IN THE 1960s

As seen in Chapters 2 and 3, the real effective exchange rate for exports had reached a realistic, and probably even undervalued, level by the late 1950s. The import rate, by contrast, remained substantially overvalued until at least 1961. After the 1964 changes, the exchange-rate system was, for all practical purposes, unified. That change undoubtedly resulted in considerable improvement in the efficiency of the exchange-rate regime compared to the 1950s.

After 1964, the restrictiveness of quantitative controls on

imports was much reduced compared with that in the 1950s, although licensing was tightened somewhat in response to payments difficulties. For exports, the real effective rate was held fairly constant by using a variety of subsidies and other export incentives. As Table 32 shows, these included internal tax exemptions, customs-duty exemptions, and interest-rate subsidies. In addition, the "wastage allowance" described in Chapter 3, and favorable treatment accorded to successful exporters, constituted sizable, if unquantifiable, additional incentives for exporting.

While the overall bias of the exchange-rate system toward exports was probably not excessive (see the estimates of ERPs for 1968 below), the use of the wastage allowance and interest-rate subsidies was a less efficient means of encouraging exports than reliance upon the exchange rate would have been. The wastage allowance was significant primarily because the import exchange rate, especially for duty-free goods, was overvalued. The effect was to encourage the use of imported inputs and to discriminate against domestic sources of raw materials and intermediate goods.[11] Nonetheless, it would have encouraged uneconomic use of imported inputs only in those instances where domestic inputs could have competed at a unified exchange rate. It is therefore doubtful whether the wastage allowance was a significant deterrent to development of intermediate-goods and raw-materials industries in Korea, although it was an important incentive for exporting. While there were undoubtedly instances of disadvantaged industries, the quantitative importance of foregone domestic output of intermediate goods and raw materials must be judged to be rather small.

Similar considerations applied to customs duty exemptions which, insofar as they only rebated duties paid on imported inputs, were really an offset to a disincentive for exports that otherwise would have been present. Even the internal tax exemptions can be viewed as a generalized incentive for export that would have had little resource-misallocation effects over and above those inherent in the bias of the trade-and-payments regime.

The same, however, cannot be said for interest-rate subsidies. For, insofar as exporters received access to subsidized credit, their incentives for using capital, and especially imported capital goods, were greater than they should have been. Interest-rate policy affected the choice of activity and technique in a variety of ways. First, there was the interest-rate subsidy that accrued directly to exporters and whose value is included in computations of effective exchange rates. Second, exporters received preferential treatment in their loan applications, and loans were made at implicitly subsidized interest rates. Third, exporters were given preferential access to foreign loans, which were guaranteed in dollars, subject to an interest rate less than half that payable on domestic borrowing. To the extent that devaluation was unanticipated or unlikely, the differential between the domestic and the foreign rate of interest led both to subsidization of loans and to distortions in the international capital market. This latter affected both the efficiency of capital flows and choice of technique, discussed below, and exporters' costs of capital. Insofar as access to foreign loans meant that the real interest rate confronting exporters was well below the opportunity cost of resources, the result was an implicit subsidy to exporters.

Wontack Hong has estimated the average interest rate paid, in nominal and real terms, on various classes of loans. His estimates of nominal and real interest rates on deposit money bank (DMB) loans and loans of the Korea Development Bank (which together accounted for somewhere between 27 and 45 percent of the financing of gross domestic capital formation in each year from 1967 to 1975) are given in Table 47, along with the Frank, Kim, and Westphal estimates of the nominal and real interest rates on foreign commercial loans (which averaged about 25 percent of GDCF).

The estimated nominal interest rates are weighted averages of all new loans in the year in question. Real rates of interest are

TABLE 47 Weighted Average Interest Rates on Loans
By Deposit Money Banks, Korea Development Bank,
and Foreign Commercial Sources, 1961–1975
(% per annum)

	DMB Loans		KDB Loans		Foreign Commercial Loans	
	Nominal	Real	Nominal	Real	Nominal	Real
1961	13.3	0.1	n.a.	n.a.	—	—
1962	13.4	4.0	8.4	-1.0	—	—
1963	13.1	-7.5	8.3	-12.3	—	—
1964	13.3	-21.3	8.4	-26.2	—	—
1965	16.2	6.2	9.2	-0.8	5.6	-2.5
1966	21.4	12.5	11.8	2.9	5.7	-2.4
1967	21.8	15.4	12.5	6.1	6.1	-2.0
1968	21.2	13.1	12.7	4.6	5.9	-2.2
1969	20.5	13.7	12.2	5.4	7.1	-1.0
1970	17.6	8.4	12.5	3.3	7.0	-1.1
1971	16.4	7.8	12.4	3.8	—	—
1972	17.7	3.7	9.9	-4.1	—	—
1973	13.9	7.0	9.7	2.8	—	—
1974	14.0	-28.1	9.7	-32.4	—	—
1975	n.a.	n.a.	11.2	-15.3	—	—

Sources: Wontack Hong, Statistical Appendix, Table 4.8 of mimeo for DMB and
KDB loans; Frank, Kim, and Westphal, p. 116, for foreign commercial loans,
1965 to 1970.

estimated by subtracting the percentage rate of increase in
prices from the nominal interest rate. To be sure, when the
expected rate of inflation was below the actual rate, borrowers
may not have perceived that the interest rate they were obtaining funds at was as favorable as it actually was, but the converse
holds for periods during which the rate of inflation was expected
to be higher than it actually was. The real rate of return
to capital in Korea must surely have exceeded 10 percent, if not
more, over the 1965–1975 decade. As can be seen, it was only

in the years 1966 to 1969 that the real rate of interest on DMB loans was anywhere at all close to that level; in other years, there was a substantial subsidy element in the real interest rate charged.[12]

It is apparent that all loan recipients were therefore implicitly subsidized when loans were granted. Given the very high priorities attached to promotion of exports, export projects and short-term export financing were among the top-priority categories for lending and for eligibility for foreign loans. For deposit money banks, for example, an increasing fraction of loans went to finance short-term export credits.[13] By 1973, new export credits exceeded the net increase in bank notes and coins issued in that year.[14] Likewise, the KDB was under instructions to accord financing of equipment and other exporters' needs top priority in allocating their funds (which included the Japanese foreign credits).

While the precise fraction of all credit that was allocated to exporters cannot be estimated, it is evident that, in conjunction with the urgency assigned by the government to export promotion, the very existence of credit rationing provided a sizable implicit subsidy to exporters. But, in addition to that, exporters paid lower nominal interest rates than those shown in the table. The availability of such credit and the preferential interest rates must have, on occasion, induced the choice of relatively more capital-intensive techniques than would have been desirable had exporters had to pay the opportunity cost of capital. Likewise, there were probably occasions when industries that were more capital-intensive than Korea's comparative advantage would have dictated embarked on exporting ventures.[15] It will be seen below, however, that this phenomenon could not have dominated the export drive.

EFFECTIVE RATES OF PROTECTION AND INCENTIVES, 1968

It was seen in Chapter 4 that there was significant liberalization of the trade-and-payments regime during the 1964–1967 period.

Thereafter, there were minor changes in the incentives provided for exports and the restrictions and charges upon imports, but these appear to have been confined to a fairly narrow range. To all intents and purposes, the salient characteristics of policy toward exports, imports, and import substitutes had stabilized by 1967, and the regime was little altered after that date. It is possible, therefore, to examine data for a single year after 1967 to evaluate the allocative efficiency of the regime.

If all policies affect prices of outputs and inputs, the best measure of the incentives confronting producers in an economy is the effective rate of protection (ERP). Appropriately measured, ERP estimates provide an indication of the degree of protection and/or subsidization accorded to different value-adding processes.[16] A higher ERP for one industry than another reflects a greater incentive for domestic production (due to a greater differential between the value added that is possible domestically and that prevailing internationally) of the more protected industry.[17] Very high ERP rates are usually symptomatic of relatively inefficient resource allocation resulting from protection, as producers are enabled to compete despite costs of production well above world levels; insofar as the same resources could be employed in other industries with much lower (and even negative) effective protection rates, significant gains in resource allocation could be achieved. This is because additional resources employed in the low-ERP industry could enable the economy to export part of that output (or import substitutes in the low-cost activity) and obtain imports of the high-ERP commodity with a considerable net saving of resources. To be sure, effective rates of protection sometimes enable domestic producers to obtain monopoly profits within a sheltered domestic market; such monopoly positions also have undesirable effects although they are not necessarily indicative of the same degree of resource misallocation.

High ERPs, and variations in them among industries, therefore, are broadly indicative of the extent of inefficiency permitted by the trade-and-payments regime, either in enabling high-cost value-adding industries to produce and bid resources away from other, more economic endeavors, or in providing domestic producers with a sheltered domestic market. High proportionate variation in ERP levels across industries is generally reflective of a trade-and-payments regime (and the presence of other incentives) that permits sizable inefficiency in the domestic production structure.

When some incentives come in the form of tax exemptions and differential incentives for foreign and domestic sales, it is necessary to take into account these incentives as well. While some estimates of ERPs take these phenomena into account, many investigators prefer to treat the "effective subsidy" as comprising the total protection implicit in tax exemptions, subsidized credit, and other inducements, along with the incentives created by the trade-and-payments regime.

For Korea, there is an excellent set of estimates of effective protective and subsidy rates for 1968 done by Westphal and Kim.[18] The Westphal and Kim computations are based, not on legal tariff rates which, as already seen, do not represent actual tariff rates due to exemptions, but on direct comparisons between domestic and foreign prices. Such comparisons enable taking into account the entire range of incentives and charges which confront producers, and are therefore vastly to be preferred to estimates of ERPs based simply on tariff tables or tariff collections.

Table 48 gives a sample of the Westphal-Kim estimates for representative industries among their 150-sector breakdown. The first column indicates the nature of the industry, and the second indicates its classification. The broad industry groups into which the individual industries belong, and the criteria for classification, are given in notes to the table. The third and fourth columns give the average nominal tariff rate and the actual amount of tariff collections. Estimates of effective rates of protection are given in the fifth and sixth columns, while the last two columns give estimates of the effective subsidy rates— taking into account both the protection as estimated in the fifth

TABLE 48 Nominal and Effective Protection Rates and Effective Subsidies,
Individual Industries, 1968
(% of international value added)

Industry	Classi-fication	Tariff	Nominal Protection	Effective Protection		Effective Subsidy	
				Exports	Domestic Market	Exports	Domestic Market
1. Rice	NIC	23.4	13.3	-0.3	14.5	2.6	19.2
11. Livestock	NIC	20.9	19.8	-1.0	11.4	7.5	10.9
12. Forest products	NIC	9.8	5.6	-0.5	4.4	0.5	4.6
13. Fishing	NIC	29.9	0.0	1.8	-4.2	11.0	1.0
17. Canned sea food	X	28.7	3.1	-3.0	11.8	9.1	4.7
23. Refined sugar	NIC	38.8	0.0	-0.3	-38.0	3.9	-43.6
32. Tungsten ore	X	0.0	0.0	-0.5	-5.8	4.4	-26.8
44. Cement	NIC	13.3	2.8	-3.7	-12.8	5.3	-11.7
47. Cotton yarn	NIC	23.1	0.0	13.7	-15.0	12.1	-18.5
48. Silk yarn	X	5.3	0.0	-2.1	-2.8	-4.5	-15.7
53. Lumber	NIC	25.7	0.0	19.0	-18.5	14.3	-26.3
54. Plywood	X	10.2	0.0	30.9	-28.4	41.8	-35.9
56. Synthetic resins and fibers	IC	40.6	24.1	-0.4	37.1	19.1	49.9
57. Petroleum products	NIC	38.3	-24.9	1.0	-66.2	2.0	-69.4

TABLE 48 (continued)

Industry	Classi-fication	Tariff	Nominal Protection	Effective Protection		Effective Subsidy	
				Exports	Domestic Market	Exports	Domestic Market
60. Glass products	IC	87.8	8.7	2.2	-4.9	3.5	-12.4
61. Pig iron	IC	9.4	14.3	-12.4	28.9	280.4	260.0
62. Steel ingots	NIC	11.4	11.4	-7.1	-12.2	-5.3	-14.8
65. Cotton fabrics	X	73.6	23.4	-8.7	169.5	93.8	176.2
66. Silk fabrics	XIC	91.6	45.2	260.6	198.8	13.0	233.8
72. Wood products	NIC	58.8	2.1	-8.5	-8.3	2.3	-10.4
74. Paper and paperboard	IC	44.9	14.8	-5.7	14.8	-7.8	2.7
76. Tires and tubes	NIC	93.9	0.0	1.2	-44.3	-14.0	-57.2
78. Basic inorganic chemicals	IC	47.5	21.6	-0.4	22.1	9.2	19.1
84. Pesticides	NIC	28.8	47.0	1.3	62.0	-8.4	58.2
86. Fertilizers	IC	0.0	4.8	-6.3	-2.9	35.1	29.0
88. Steel sheets and bars	IC	24.4	27.8	-7.5	138.7	15.0	-3186.6
94. Tools and other metal products	XIC	36.2	27.1	-4.3	57.8	4.4	55.0
98. Knit products	X	61.3	11.7	-2.4	31.0	2.3	14.0
104. Leather shoes	X	89.9	6.7	-1.7	-3.0	5.2	-9.9

TABLE 48 (continued)

Industry	Classification	Tariff	Nominal Protection	Effective Protection		Effective Subsidy	
				Exports	Domestic Market	Exports	Domestic Market
110. Pottery	IC	76.4	49.6	-2.5	97.3	52.3	96.3
117. Toys and sporting goods	XIC	74.9	4.2	-2.4	-14.3	9.5	-21.1
121. Electronic components	XIC	1.6	1.6	-2.1	39.2	6.3	49.6
126. Bicycles	NIC	96.6	20.5	-3.5	4.7	0.4	-4.9
130. Metal working machinery	IC	20.9	4.5	-3.3	-10.6	-5.3	-14.7
134. Textile machinery	IC	13.9	0.5	-2.8	-16.8	-0.1	-16.0
137. Office machines	XIC	70.4	24.2	-4.1	34.2	7.3	31.9
140. Generators	NIC	8.6	5.2	-3.1	-9.2	19.5	4.8
144. Electronic equipment	IC	51.7	44.6	-5.8	64.5	0.7	48.1
149. Motor vehicles	IC	121.5	88.0	-13.5	247.7	-6.1	241.8

Source: Westphal and Kim, "Industrial Policy and Development," Appendix Tables 2A and 2B.

Notes: a. Industries 1–13 are designated agriculture, forestry and fishing; 14–25 are processed foods; 26–29 beverages and tobacco; 30–43 mining and energy; 44–46, construction materials; 47–64 intermediate products I; 65–97 intermediate products II; 98–121 non-durable consumer goods; 122–127 consumer durables; 128–145 machinery; and 146–150, transport equipment.

b. X = export industry (exports greater than 10% of total production)
  IC = import-competing industry (imports greater than 10% of total domestic supply)
  XIC = both X and IC
  NIC = non-import competing industry (all industries with less than 10% exported and less than 10% of total supply imported)

c. Westphal and Kim give both Balassa and Corden estimates of effective protection and subsidy rates. The Corden measures are reproduced here.

and sixth columns, and divergences of tax liabilities and interest rates paid from the average for all industry.

Several features are noteworthy. First, the "true" nominal protection afforded to various industries does not, in general, appear to have been exceptionally high. Judged by the standards of most other countries, nominal protective rates in Korea on the whole seem fairly moderate, despite some fairly high legal tariff rates, when exemptions, and so on, are not taken into account. Second, in most instances, effective protective rates appear to be even more moderate compared to levels found elsewhere. Here, of course, account must be taken of the difference between selling prices in the domestic market—some of which were considerably above prices of comparable products for export—and for export. In general, effective protective rates for export are mildly negative, while those for the domestic market range from minus 44 percent (for tires and tubes) to a positive 247 percent (for motor vehicles). The silk fabrics industry, which is both exporting and import-competing, is the only industry with very high rates of effective protection both for export and for the domestic market.

When attention turns to effective subsidy rates, however, the picture changes rather dramatically. In particular, most industries were confronted with positive rates of effective subsidy when selling in the export market, and most had greater total incentives to export than to sell domestically. Exceptions seem to lie mostly in the import-competing industries, such as pottery and basic inorganic chemicals, where the effective subsidy rate for the domestic market exceeds that for export. To be sure, there are anomalies, such as cotton fabrics, which apparently received fairly high effective subsidies both for the domestic market and for export.

In general, the thrust of Korea's export-promotion policy shows through clearly. Unlike most other countries, where effective rates of protection and subsidy are systematically and significantly higher for import substitutes than for export, in Korea export effective subsidy rates in general exceeded those for the domestic market. In this regard, the importance of the credit and tax incentives shows through clearly: the effective rates of protection by themselves usually did not provide more incentive for exports than for the domestic market.

Some summary statistics serve to highlight the wealth of information contained in the detailed ERP and effective subsidy rates calculated by Westphal and Kim. Table 49 gives estimates of nominal and effective protective rates and effective subsidy rates by industry group. These estimates were made in the same way as the data given in Table 48, and then weighted to form industry aggregates. As can be seen, the structure of Korean protection was unusual in a number of respects. Unlike most other countries, the average nominal protection to primary commodities exceeded that accorded to manufacturing. Thus, for a large number of manufacturing sectors, effective rates of protection were negative. When account is taken of differential tax treatment and credit subsidies, however, export effective subsidies once again are positive (for all manufacturing industries except Transport Equipment), while a number of industries faced negative effective rates when selling on the domestic market. Intermediate Products I—which includes such products as cotton yarn, lumber, plywood, petroleum products, glass products, pig iron, steel ingots, copper, and other nonferrous metals—is the category with the largest differential incentive between export and domestic sales, with an effective subsidy rate of plus 26 percent for export and minus 22 percent for the domestic market. Transport Equipment consisted almost entirely of import-competing activities in 1968, which probably accounts for the unusual combination of high incentives for the domestic market and an absence of incentives for export.

In addition to the differential in incentives for domestic and foreign sales, the other notable characteristic of Table 49 is the relatively moderate levels of protection and effective subsidies. Except for Transport Equipment, no sector received an effective subsidy in excess of 26 percent, and only three manufacturing sectors benefited from effective subsidies for export in excess of

TABLE 49 Nominal and Effective Protective Rates and Effective Subsidies, Industry Group, 1968 (%)

Industry Group	Average Actual Tariff	Average Nominal Protection	Effective Protection		Effective Subsidy	
			Export	Domestic Sales	Export	Domestic Sales
Agriculture, Forestry and Fishing	36.0	16.6	-15.3	17.9	-9.4	21.7
Mining and Energy	9.6	6.9	-0.9	3.5	2.7	4.5
Total Primary Production	34.1	15.9	-7.0	17.1	-2.4	20.7
Processed Foods	56.7	2.7	-2.2	-14.2	1.8	-19.6
Beverages and Tobacco	135.4	2.1	-1.7	-15.5	12.6	-20.8
Construction Materials	30.5	3.7	-3.9	-8.8	4.4	-12.9
Intermediate Products - I	31.0	2.4	18.6	-18.8	26.0	-21.9
Intermediate Products - II	53.4	19.1	-0.2	17.4	11.6	13.1
Nondurable Consumer Goods	67.9	8.6	-1.4	-8.0	4.1	-15.7
Consumer Durables	78.4	30.7	-3.0	39.8	1.5	23.6
Machinery	49.1	27.9	-4.6	29.5	1.9	21.0
Transport Equipment	61.8	54.3	-13.1	83.2	-5.6	80.8
TOTAL MANUFACTURING	58.8	10.7	2.2	-1.1	8.9	-6.5
ALL INDUSTRIES	49.4	12.6	0.3	9.0	6.5	8.6

Source: Westphal and Kim, "Industrial Policy and Development," Tables 2.A and 2.B.

10 percent. The range of effective subsidies was somewhat greater for the domestic market, presumably reflecting the dual motivation of the government in providing protection: on the one hand, there were industries where the motive was primarily to encourage export sales, so that the domestic market ended up with negative effective subsidy rates; on the other hand, there were sectors where import-substitution industries predominated, and in those cases, positive subsidy levels reflect the government's desire to encourage those sectors.

These conclusions are reinforced by the summary statistics in Table 50, again drawn from Westphal and Kim's study and presenting the same statistics as Table 49, except that the aggregation is by sales destination of commodity. Export industries, that is, those which sold more than 10 percent of their output in the export market, were on average provided with an effective subsidy rate of about 10 percent for foreign sales in manufacturing, while sales to the domestic market were effectively taxed about 20 percent. Manufacturing import-competing industries, by contrast, received effective subsidies of about 16 percent for the export market, and 50 percent for the domestic market. Non-import-competing activities were accorded relatively low levels of protection, while the industries in which both export- and import-competing sectors were important received positive effective subsidies for both destinations. In general, incentives to exports appear to have been fairly uniform whereas import-substitution incentives were far more selectively provided.

EXPORT POLICY

If one uses the Westphal-Kim estimates of effective subsidy as a guide, the average effective subsidy to manufactured exports was about 9 percent, while that to manufactures for the domestic market was minus 6 percent. This would imply a 15 percent differential in favor of exports, compared with what incentives would have been in the absence of government intervention.

TABLE 50 Nominal and Effective Protective Rates and Effective Subsidies, Commodity Categories, 1968
(%)

Commodity Category	Average Actual Tariff	Average Nominal Protection	Effective Protection		Effective Subsidy	
			Export	Domestic Sales	Export	Domestic Sales
A. Export Industries						
Primary Activities	1.4	-7.6	-11.6	-12.0	-8.5	-19.0
Manufacturing	53.7	5.2	3.4	-14.0	9.8	-20.4
Total	51.2	4.6	0.7	-13.9	6.5	-20.3
B. Import-Competing Activities						
Primary Activities	22.8	46.3	-0.2	66.6	4.6	74.2
Manufacturing	55.4	31.6	-3.9	55.1	15.8	50.2
Total	47.5	35.2	-3.6	57.3	15.0	59.8
C. Non-Import-Competing Activities						
Primary Activities	36.2	13.3	0.6	12.7	8.2	16.1
Manufacturing	64.1	5.0	-0.7	-12.6	5.0	-18.7
Total	49.5	9.3	-0.1	3.8	6.5	3.9

TABLE 50 (continued)

Commodity Category	Average Actual Tariff	Average Nominal Protection	Effective Protection		Effective Subsidy	
			Export	Domestic Sales	Export	Domestic Sales
D. Export- and Import-Competing Industries						
Primary Activities	1.2	7.6	-1.3	10.6	2.6	13.7
Manufacturing	46.3	23.1	-1.4	46.1	5.6	34.8
Total	44.3	22.5	-1.3	43.6	5.3	33.4

Source: Westphal and Kim, "Industrial Policy and Development," Tables 3A and 3B. See Table 48 for definition of commodity categories.

This estimate bears out the impression gained from inspection of export-promotion policies—that the totality of government policy was oriented toward export promotion, and that incentives were biased somewhat toward exports and against the domestic market. However, examination of effective incentive rates and other evidence does not bear out the allegation sometimes heard that the Korean regime was as chaotic and indiscriminate in its export orientation as other regimes have been toward import substitution. Since policies toward exports were one of the key areas in Korea's growth strategy, it is important to provide reasons why the "indiscriminate" export promotion hypothesis does not withstand close scrutiny.

First, much of what discrimination did exist was not between exports and import substitutes; it was between sales to the home market and sales abroad. This is most clearly seen by examining the contrast between effective subsidy rates for domestic sales and for exports; discrimination came at the point of sale, not at the point of production. Second, and closely related to the first, Korean government strategy usually discriminated in favor of exports and not in favor of specific commodities. By and large, the resulting differentials in effective subsidy rates were not the result of conscious government policy favoring one sort of export over another. Thus, any firm that exported was eligible for export incentives; many of the incentives were significant only if the firm could profitably export. To the extent that that was so, a market test still retained importance in selecting appropriate export industries, since prices provided one important determinant of profitability, and the incentives were effective only to the extent that profits were realized.

To be sure, such a link was not perfect, and probably some of the export incentives resulted in less than an optimal mix of exports. Nonetheless, the ERP estimates, as well as other data, all suggest that divergences in the output mix were of limited magnitude.

If incentives did not discriminate unduly among exports, they surely encouraged techniques of greater-than-optimal capital intensity, especially in the early 1970s. Several pieces of evidence are available. First, there are two sets of estimates of the factor intensity of exports. In addition, Rhee and Westphal have made a careful microeconomic estimate of the impact of the incentive structure on the choice of technique.

The two sets of estimates of the factor intensity of Korean trade are those of Westphal and Kim and of Wontack Hong. The Westphal-Kim estimates are based entirely on 1965 and 1966 input-output coefficients and factor inputs (labor and capital at 1968 prices) per unit of output. These data, while highly reliable, do not take into account changes in factor intensity within industries that might have occurred over the period covered by their estimates, 1960 to 1968. They do, however, reflect the changing commodity composition of trade. Westphal and Kim estimated the direct and direct-plus-indirect labor and capital requirements for exports and import-competing goods for primary commodities and manufactures separately.

Wontack Hong's data are not entirely comparable for a number of reasons. First, he used annual data in constructing his estimates, so that his coefficients refer to the year of the estimate and therefore include both changes in the output mix and changes in factor intensity within industries. Second, Hong's estimates are in 1970 prices, contrasted with 1965 prices for the Westphal-Kim data. Third, by virtue of a different base year, the commodity categories used by the two are not entirely comparable. Finally, Hong's estimates for direct-plus-indirect inputs follow the Corden concept—that is, they include indirect home goods only.

The results of the two sets of estimates are presented in Table 51. The Westphal-Kim estimates clearly show the importance of treating primary and manufacturing industries separately. The direct labor requirements for manufacturing exports increased slightly from 1960 to 1968 according to the Westphal-Kim estimates while capital requirements for exports fell. Thus, the labor-capital ratio for manufactured exports rose

TABLE 51 Estimates of Factor Intensity of Production, by Commodity Categories

	Westphal-Kim Estimates, 1960 to 1968							
	1960			1963	1966	1968		
Direct Factor Requirements	L	K	L/K	L/K	L/K	L/K	L	K
Primary								
Domestic Output	10.86	0.65	16.60	17.20	17.08	17.16	10.74	0.63
Exports	7.54	0.92	8.19	6.89	6.15	5.69	6.27	1.10
Imports	11.06	0.67	16.58	15.91	16.13	15.48	11.28	0.73
Manufacturing								
Domestic Output	1.63	0.55	2.97	2.89	2.67	2.64	1.53	0.58
Exports	1.87	0.69	2.72	3.02	3.24	3.55	1.89	0.53
Imports	1.29	0.62	2.09	1.93	1.98	2.33	1.54	0.66
Total								
Domestic Output	5.44	1.24	4.39	4.59	4.46	4.12	4.48	1.09
Exports	4.83	1.49	2.52	2.52	2.41	2.10	2.56	1.22
Imports	3.37	0.74	4.87	4.87	4.05	4.29	2.96	0.70
Total Factor Requirements								
Primary								
Final demand less imports	12.86	1.12	11.46	11.79	12.10	12.61	13.36	1.06
Exports	9.84	1.49	6.55	5.75	5.13	4.81	8.29	1.73

TABLE 51 (continued)

Westphal-Kim Estimates, 1960 to 1968

Total Factor Requirements	1960 L	1960 K	1960 L/K	1963 L/K	1966 L/K	1968 L/K	1968 L	1968 K
Primary (continued)								
Imports	12.99	1.08	11.99	11.50	11.90	11.30	13.06	1.16
Manufacturing								
Final demand less imports	8.92	1.64	5.43	5.41	5.03	5.14	8.53	1.66
Exports	7.89	2.11	3.74	3.71	4.09	4.29	7.91	1.83
Imports	5.06	1.84	2.77	2.40	2.40	2.74	5.56	2.03
Total								
Final demand less imports	9.50	2.16	4.39	4.59	4.46	4.12	9.32	2.26
Exports	8.12	2.38	3.42	3.05	3.25	3.15	7.53	2.38
Imports	6.74	1.79	3.78	3.66	3.26	3.48	6.62	1.89

Wontack Hong's Estimates

	1960	1963	1966	1968	1970	1973
Capital Requirements (million 1970 dollars per $100 million exports or import replacements)						
Direct Exports	43	48	41	41	41	44
Indirect Exports	50	68	49	50	58	44
Total Exports	93	116	90	91	98	88

TABLE 51 (continued)

			Wontack Hong's Estimates (continued)			
	1960	*1963*	*1966*	*1968*	*1970*	*1973*
Capital Requirements (continued)						
Direct Imports	48	41	58	49	45	44
Indirect Imports	59	58	83	69	70	77
Total Imports	107	99	141	118	115	121
Labor Requirements (1,000 workers per $100 million exports or import replacements)						
Direct Exports	105	65	59	49	39	23
Indirect Exports	50	71	38	32	27	17
Total Exports	155	136	97	81	66	40
Direct Imports	45	54	40	41	43	34
Indirect Imports	47	45	39	37	28	21
Total Imports	92	99	79	78	71	55
K/L Ratio (1,000 dollar capital per worker)						
Direct Exports	.41	.74	.69	.84	1.05	1.91
Indirect Exports	1.00	.96	1.29	1.56	2.15	2.59
Total Exports	.59	.85	.93	1.12	1.48	2.20
All Manufacturing	n.a.	1.53	1.53	1.44	1.67	1.58
Direct Imports	1.07	.76	1.45	1.20	1.05	1.29

TABLE 51 (continued)

	Wontack Hong's Estimates (continued)					
	1960	*1963*	*1966*	*1968*	*1970*	*1973*
K/L Ratio (continued)						
Indirect Imports	1.26	1.29	2.13	1.86	2.50	3.69
Total Imports	1.16	1.00	1.78	1.51	1.62	2.20

Sources: Westphal and Kim, "Industrial Policy and Development"; Wontack Hong, "Trade, Distortions and Employment," Table 7.10 (revised edition 1977).

Notes: a. Hong's estimates are for all sectors, primary included. They therefore correspond with the Westphal-Kim estimates given under "total."
b. Hong's estimate for 1973 was derived using 1970 input coefficients.

from 2.72 to 3.55 according to their estimates. Direct factor requirements for manufactured exports were, therefore, below those for domestic output but above those for import substitutes in 1960; by 1968, manufactured exports had a higher labor-capital ratio than either domestic output or import substitutes.

These findings are obscured in the estimates of factor requirements for all commodity categories because primary commodities were apparently highly labor intensive for import substitutes compared to the coefficients for exports. As a consequence, total exports were less labor intensive than either domestic output or imports throughout the period. When direct and indirect factor requirements were computed, the results were not significantly affected: the labor-capital ratio in manufacturing rose throughout the period (except for 1963), while that for imports declined from 1960 to 1966, and rose again thereafter. When primary commodities and manufacturers are grouped together, the labor intensity of primary-commodity import substitutes dominates the figures, thus leaving imports more labor-intensive throughout.[19]

Hong's estimates span a longer time period than the Westphal-Kim data, but are not available for manufacturing and primary commodities separately. His data (Table 51) show direct labor requirements for exports falling from 1960 onward, direct capital requirements staying constant, and hence a rising capital-labor ratio in exports throughout, with the exception of 1963, a year that appears to have certain anomalies in both sets of estimates. Hong's estimates show direct requirements for exports to be more labor-intensive than those for imports, and total exports (direct-plus-indirect) are more labor-intensive than total imports until 1970.

The picture that seems to emerge is that the beginning of the Korean export boom from 1960 to 1970 was a period during which the exports that grew rapidly were of greater labor intensity than was Korean manufacturing as a whole. Simultaneously, the capital-labor ratio for import substitutes was

either constant, or rising very slowly. One possible interpreta-
tion is that, during this period, the factor content of exports and
import substitutes was growing closer together, perhaps because
capital-intensive industries that had been started in the 1950s
diminished in importance, and perhaps in response to the failure
of the real wage to rise as unemployed labor was being absorbed
while the real interest rate increased substantially.[20]

Since Korean comparative advantage at that time surely lay in
labor-intensive industries, the Korean experience is consistent
with the notion that the export-promotion policies of the 1960s
resulted in resource reallocation toward the industries with
comparative advantage. This was probably true both for export
and for import-competing industries.

After about 1968, rapid capital accumulation resulted in
a gradual change in comparative advantage toward less labor-
intensive commodities. That export and import-competing
industries both became more capital-using is consistent with
that. Moreover, capital-labor ratios in import-competing and
export industries moved closer together over the decade. It
seems clear that ideal resource allocation would, if capital and
labor were the only factors of production, dictate approximate
equality of those ratios. All of these considerations, as well as
the relatively small variation in effective protection and subsidies
compared with other LDCs, suggest that resource allocation
improved significantly during the 1960s.

After 1970, increasing capital intensity, as shown in Wontack
Hong's estimates, is broadly consistent with the rising wage-
rental ratio that characterized the economy in those years. To
the extent that exports were becoming more capital-intensive
than import substitutes, it is possible that interest-rate subsidies, which were of increasing importance both proportionately and absolutely in the early 1970s, began inducing even
more capital-intensive exports and techniques than were optimal.

A partial explanation of the increasing capital-intensity of
exports, as revealed in Hong's data, however, may lie in the
fact that some of the import-substitution industries started up

in Korea during that time: steel, fertilizer, and petrochemical derivatives. As Westphal and Kim argue,

> All of these products require capital-intensive production methods in plants subject to severe economies of scale . . . Given a decision to meet the domestic demand for these commodities through domestic production, temporary exports can be efficient as they permit the construction of large plants without experiencing the initial excess capacity . . . Even without such a decision, exports of cement, steel, and fertilizer during the first half of the 1970's may well have been in Korea's dynamic comparative advantage.[21]

Westphal and Kim, however, appear to question the wisdom of the decision to produce petrochemicals to supply the textile and plastic industries.

Thus, part of the increased capital intensity of exports reflected in Hong's data reflects altered comparative advantage. Part reflects efficient utilization of large-scale plants in import-substitution industries (which may or may not themselves have been efficient), and part undoubtedly reflects the use of overly capital-intensive techniques induced by credit subsidies as a form of export incentive. Moreover, even this source of non-optimality should be kept in perspective: there can be little doubt that the Korean economy of the 1970s had a vastly more efficient resource allocation than that of the early 1960s. The point is that the export drive itself was relatively efficient, although there were some deviations from optimality.

A study by Rhee and Westphal[22] tends to confirm the conclusion that microeconomic inefficiencies, in the form of non-optimal capital intensity, were present in the early 1970s. They examined the capital equipment purchased for cotton-textile weaving in the period 1970 to 1973. It should be noted that cotton-textile weaving was one of the least effective industries in Korea, judged by the height of effective protection granted to it in 1968, which serves to put the finding of inefficiency in perspective.

In the cotton-textile-weaving industry, producers had two

choices: domestically produced semi-automatic looms, and imported automatic looms. The purchase price of a domestically produced semi-automatic loom was about 30 percent of the price of an imported one. Domestic looms were, naturally, less capital- and more labor-using. Nonetheless, large numbers of imported looms were purchased, and Rhee and Westphal set out to find out why. Interviewing 79 textile firms (with 233 models of looms), they estimated the technological relationships pertaining to each loom.

They then estimated the benefits and costs that would accrue to each firm as a function of the type of loom purchased, taking into account as many microeconomic characteristics of the product and the loom alternatives as possible. Their conclusions are worth quoting at some length:

> It is a fair generalization to say that large producers have monopolized the production of luxury fabrics where high profits are to be made through export subsidies and discriminatory pricing on the protected domestic market. One might wonder why these producers export at all, but exporting is the price paid to do business in Korea ... In turn, large scale producers tend to be inefficient, because of their reliance on imported automatic technology. The only clear exception to this is in the production of wide cloth. But the inefficiency of large scale producers appears not to result from any failure to maximize profits on their part, it is rather due to government policies which have favored capital imports.[23]

Rhee and Westphal also noted that smaller-scale producers, not able to benefit from government subsidies, were efficient producers using domestic capital equipment. Being ineligible for export subsidies, however, they exported relatively little.

From the myriad details underlying the Rhee-Westphal estimates it is apparent why more microeconomic studies are not available. Nonetheless, their conclusions are significant in confirming both that the qualitative predictions of economic theory are in fact borne out and that producers, in responding to incentives, were using overly capital-intensive techniques of production. It should be noted that the incentives that resulted in the use of overly capital-intensive techniques were removed after 1972, which is consistent with the view that reduced bias toward exports after 1972–1973 probably was economically justified.

The overall picture, therefore, seems to be one of increasing efficiency of resource allocation—both in the commodity composition of output and in factor proportions—throughout the period after 1960. From 1960 to at least 1968, the improvement in resource allocation must have been very significant. Thereafter, opportunities for further gains were not as great. In the 1970s, comparative advantage was shifting, but there is some evidence that the shift toward capital-using techniques was somewhat too rapid. There can nonetheless be little doubt that the efficiency of trade in 1975 far exceeded that of the 1950s.[24] Indeed, one of the factors responsible for rapid growth under export promotion must certainly have been the improved resource allocation that resulted from it and the other policies that accompanied the export-oriented strategy.

CAPITAL FLOWS

It has already been seen that there were virtually no capital flows other than foreign aid prior to the 1960s. From 1960 to 1965, private capital flows were relatively small, and did not play a significant part in resource allocation. This was not a consequence of policy failure: it is undoubtedly true that Korea could not have begun to attract significant non-concessionary private capital until such time as her export-promotion policies had begun exhibiting success. For that reason, it is policy only with respect to private capital flows after 1965 that needs to be evaluated here.

As with other aspects of resource allocation, it is likely that the capital inflows were economically justified. There can be little doubt that the social rate of return on capital exceeded the foreign borrowing rate, so that the inflow of funds was generally well used. Nonetheless, there were incentives to borrow excessively, and some misallocation undoubtedly resulted, especially in the 1968-1971 period. After that, these incentives diminished.

The divergence between the real rate of interest as perceived by businessmen and the real social cost of borrowing was described in Chapter 4. It will be recalled that, during the late 1960s, exchange-rate depreciation was limited and well below the domestic rate of inflation. All borrowing from abroad for maturities in excess of one year was denominated in foreign currencies. Thus, a Korean businessman borrowing from abroad could reasonably expect his revenues—even if he was an exporter, since export subsidy policies could be expected to maintain the real exchange rate for him in the absence of devaluation—to keep pace with the domestic rate of inflation and the exchange rate to lag well behind. From 1965 to 1970, the average rate of increase of the GDP deflator was 11.3 percent, and the average rate of depreciation of the currency was 3.2 percent. With interest rates on foreign loans ranging between 6 and 8 percent, it is evident that the majority of borrowers paid negative real interest rates on their loans. Since inflation had been at even higher rates in earlier years, it is likely that rates of inflation in excess of the interest rate were anticipated.

Had all who wished to do so been free to borrow at the foreign interest rate, there is little question but that the aggregate level of foreign borrowing would have been much too high to be optimal from the viewpoint of the Korean economy, with consequent resource misallocation. This would have occurred both because borrowers used more capital-intensive techniques and invested in more capital-intensive industries than they would have found profitable at the real foreign rate of interest. In addition to an above-optimal total amount of borrowing, there would have been sectoral inefficiencies, since some sectors of the economy were relatively favored and some were disadvantaged, contrasted with what would have happened had all been free to borrow at the true social cost.

In practice, however, the government did not permit foreign loans without approving them, so that implicitly subsidized loans were available subject only to rationing on the part of the government. This constrained the level of foreign borrowing well below what would have occurred had there been no quantitative intervention.[25] Moreover, the government generally permitted foreign loans to finance only a fraction of the total cost of proposed projects; loans from commercial banks and borrowing from the curb market both generally had to be used to finance a proposed project. The interest rate in fact paid by a prospective borrower was thus a weighted average of the interest rate paid to each lender. As such, it was above the nominal interest rate to be paid on foreign loans.

Nonetheless, there undoubtedly was some excessive borrowing, and some resulting resource misallocation, in the late 1960s.[26] Frank, Kim, and Westphal report that domestic borrowers sharply reduced their demand for foreign loans after the large and unexpected devaluation of 1971. In their words,

> Interviews with businessmen suggest that . . . there was no expectation that the exchange rate would change as much as it did during the late 1960's. If this is true, the large influx of foreign capital may have been due in part to an underestimate of the real private costs because of an expectation of a stable exchange rate . . . The value of the wŏn, however, gradually fell between the beginning of 1968 and mid-1971, at which time there was a sharp devaluation. Nevertheless, during 1968 and 1969, foreign commerical borrowing continued to grow sharply. In 1970, however, the demand for foreign loans was reduced sharply. Perhaps by 1970, it had become clear to businessmen that movement in the value of the wŏn was not temporary and that the true cost of foreign borrowing was likely to be greater than they had originally expected, although government ceilings on foreign borrowing may have been chiefly responsible for the slow growth of foreign borrowing in 1970.
>
> In 1971 and 1972, also, the demand for foreign commercial borrowing seems to have slackened. According to businessmen interviewed, their desire for foreign loans was curbed by the devaluation of June

1971 and by the reintroduction of the rapidly sliding peg in early 1972.[27]

Although borrowing in the late 1960s almost certainly exceeded the optimal level, it is difficult to believe that the low private real rate of interest caused quantitatively significant distortions within the economy. First, as pointed out by Frank, Kim, and Westphal, the exchange rate began depreciating in 1968. Second, the government rationed credit and did so in accordance with export performance. Third, insofar as credit rationing discriminated against any industries within Korea, it was against domestic-machinery and capital-equipment industries. But those industries, during the time when the real foreign interest rate was distorted, were not usually industries in which Korea appeared to have a comparative advantage. Fourth, as the Westphal-Kim and Hong data on factor intensity indicate, exports were concentrated in labor-intensive industries—the ones permitted to borrow abroad. Finally, part of the value of the implicit credit subsidization was included in the Westphal-Kim estimates of effective subsidies reported above. The fact that the range of effective subsidies seems fairly moderate is further reflection of the fact that foreign borrowing, although potentially dangerous, was sufficiently contained by credit rationing so that, while above the optimum, it probably did not constitute a major distortion in resource allocation.

CONCLUSIONS

Economists' prescriptions for optimal resource allocation are never exactly met in reality, and judgment must always be used in assessing the extent to which misallocations result in significant losses. In the Korean case, there is ample evidence on which to base the conclusion that large and significant inefficiencies in the trade-and-payments regime stemmed from the currency overvaluation and multiple exchange-rate practices of the 1950s. The exchange-rate regime was sufficiently unrealistic and chaotic so that important losses resulted and these were the most significant source of misallocation in the 1950s.

In terms of the industries that developed during the 1950s, however, the evidence suggests that resource misallocation, while substantial, was not as large as that which seems to have accompanied import-substitution drives in some other countries. To be sure, growth seems to have faltered in the late 1950s as the "easy" import-substitution stage was over, and that hesitation may be credited as a cost of the import-substitution policies. Nonetheless, insufficient time had elapsed for Korea to develop the entire range of intermediate-goods and capital-goods industries that would have made the transition to the 1960s much more difficult. It was thus the inefficiency of the trade-and-payments regime more than the inefficiency of the import-substitution policies that in part accounted for the significant misallocations of the 1950s.

With regard to aid, some of the contributions at the microeconomic level were fundamental—education and land reform among them. Moreover, given the importance of aid in financing the import bill, it is difficult to disassociate aid allocations from the overall allocation of resources within the Korean economy of the 1950s. Aid was sufficiently sizable that it is difficult, if not impossible, to assess its role in microeconomic terms. The aid role, at the macroeconomic level, is discussed in Chapter 6.

As for the 1960s and early 1970s, all the evidence points to the conclusion that the trade-and-payments regime, and the export-promotion policies of that era, were considerably more efficient in microeconomic terms than the policies of the 1950s. To be sure, effective subsidies did depart from uniformity, which is the economist's prescription for optimal resource allocation. Nonetheless, the range of variation in incentives for production of different commodities appears to have been reasonably narrow, as much discrimination was really between domestic and foreign sales, regardless of the nature of the commodity. If there is an identifiable source of resource misallocation in the late 1960s and early 1970s, it probably lies in the credit rationing and interest rate subsidization policies that were increasingly, as time progressed, relied upon to encourage exports. The available evidence is not sufficient to enable a definitive conclusion with regard to the probable quantitative importance of those subsidies, but it would appear that, at least until 1975, those policies did not result in significant costly misallocation. This same conclusion applies to foreign borrowing, which probably was excessive in the late 1960s, owing to the low real interest rate paid by domestic borrowers; the quantitative importance of the excess was probably not large. The overall verdict must be that the Korean export-promotion strategy of the late 1960s was generally based on relatively efficient patterns of resource allocation, and broadly reflected Korea's comparative advantage.

SIX

Macroeconomic Effects of Trade and Aid

Any distinction between macro- and microeconomic aspects of growth is tenuous, and never more so than in the case of Korea. Growth itself is a macroeconomic variable, but who can doubt that improved microeconomic efficiency, as discussed in Chapter 5, was a major contributor to the increased growth rate of the 1960s? Thus, the topics already discussed in earlier chapters have had significant macroeconomic effects and are not dealt with again here. Focus instead is on the more traditional macro-economic variables—components of GNP, and growth rates. There are three major avenues through which trade, the payments regime, and aid affected macroeconomic performance. First, there is the contribution of aid, and later capital inflows, to the total resources available for capital accumulation. A second major impact of the trade-and-payments regime, especially after the policy reforms of the early 1960s, was the effect rapid growth of exports had on the domestic economy. Third and finally, there is the impact of the trade strategy, and especially the growth of exports, on employment. These three aspects are treated in turn here. A final section attempts an overall assessment of the contribution of trade, aid, and capital flows to Korea's growth.

AID AND CAPITAL FLOWS
AS A SOURCE OF SAVING

Table 52 gives data on domestic and foreign saving for the period 1954 to 1975. The data virtually speak for themselves: net borrowing was negligible in the 1950s, and transfers consisted almost exclusively of aid. Aid contributed more than half the total resources available for capital accumulation in every year from 1955 to 1962, and in some years its contribution was substantially more than that. Indeed, in 1956, net transfers from abroad exceeded total saving, as government dissaving more than offset private domestic saving. The relative importance of aid began declining rapidly after 1962,[1] although net transfers still exceeded 40 percent of total saving in 1965. Thereafter, they declined rapidly in importance, dropping below 25 percent of total saving in 1966 and 10 percent in 1971. However, except by the standards of the earlier contribution of aid to savings in Korea, net transfers constituted a significant augmentation of total resources available for savings at least through 1970. Indeed, with gross investment equal to 24.8 percent of GNP in 1970, net transfers, which constituted 35 percent of total savings, still equaled almost 2 percent of Korean GNP; in the 1950s, of course, the figure had exceeded 10 percent.[2] Even in 1975, foreign saving financed investments equal to 10 percent of GNP, although most of that came from foreign loans.

These orders of magnitude are, by themselves, sufficient to indicate the importance of aid. First, in regard to the 1950s, a number of observations are in order. In the absence of aid, it is

TABLE 52 Domestic and Foreign Saving, 1954-1975

| | Domestic Saving | | Foreign Saving | | Total Saving[b] | Percentages | |
	Private	Government	Net Transfers	Net Borrowing		Foreign Saving of Total Saving	Gross Investment to GNP
			billions of won				
1954	6.06	-1.80	2.82	0.70	7.78	45.2	11.1
1955	8.37	-2.69	6.49	1.64	13.81	58.9	11.1
1956	2.35	-4.42	17.62	-1.14	14.41	114.4	8.5
1957	16.94	-6.01	19.24	0.09	30.26	63.9	13.9
1958	16.70	-6.43	18.65	-2.19	26.73	61.6	12.0
1959	14.60	-5.94	15.89	-0.83	23.72	63.5	10.0
1960	8.55	-5.01	22.06	-1.07	26.80	78.3	10.0
1961	16.88	-5.30	29.51	-4.22	38.79	65.2	12.0
1962	10.34	-4.86	30.73	7.22	45.47	83.4	11.8
1963	31.81	-1.32	33.73	18.63	90.26	58.0	16.7
1964	48.39	3.55	44.03	5.10	102.24	48.1	13.6
1965	46.48	14.02	53.95	-2.42	121.98	42.2	14.2
1966	93.37	29.08	59.58	28.05	224.48	39.0	20.0
1967	99.96	51.85	60.94	51.92	280.97	40.2	20.3
1968	117.71	100.61	62.54	121.79	427.87	43.1	24.0

TABLE 52 (continued)

| | Domestic Saving | | Foreign Saving | | Total | Percentages | |
	Private	Government	Net Transfers	Net Borrowing	Saving[b]	Foreign Saving of Total Saving	Gross Investment to GNP
			billions of won				
1969	235.63	129.55	70.86	158.16	620.70	36.9	26.9
1970	243.20	180.00	55.96	193.35	704.66	35.4	24.8
1971	268.17	190.10	59.32	294.68	805.35	44.0	23.0
1972	427.74	149.51	66.71	148.32	805.48	26.7	19.8
1973	864.68	225.09	75.74	123.18	1292.29	15.4	25.2
1974	1099.90	202.98	90.37	827.35	2125.88	43.2	27.6
1975[a]	1299.57	336.61	106.48	922.01	2459.78	41.8	24.4

Source: BOK, *Economic Statistics Yearbook, 1974,* pp. 300–301, *1976,* pp. 300–301.

Notes: [a]All 1975 data are preliminary.

[b]Figures do not sum to total due to statistical discrepancy.

clear that, if domestic saving had not altered, net investment would probably have been negative in several years. While it is true that depreciation rates were undoubtedly very low in the immediate post-war years, replacement investment nonetheless would have required some resources. Domestic savings as a percentage of GNP were 6.1, 4.6, minus 5.0, plus 4.6, and 3.6 percent in the years from 1954 to 1959, respectively. While these rates might have been adequate to maintain capital stock, such maintenance was from a war-torn base, and certainly would not have been sufficient to keep per capita incomes constant.

It follows, therefore, that, given domestic savings rates, aid flows in the 1950s were necessary in order to permit such economic growth and recovery as took place. By and large, that conclusion holds even if one takes into account the fact that domestic savings would probably have increased somewhat had aid flows been significantly reduced. As inspection of Table 52 shows, government saving was negative throughout the 1950s, and continued so until 1964. Such a large deficit would probably have been politically intolerable (and unsustainable, depending as it did upon the availability of foreign aid for its financing) in the absence of aid flows, or even in the presence of significantly smaller ones.

Part of the reason for this was pointed out by Cole and Lyman:

> Thus, in Rhee's time, the Korean government followed a set of policies that clearly kept the internal and external financial gaps wide open to facilitate financial and real-resource transfers from abroad and to help justify the need for more aid. These policies consisted of an overvalued exchange rate, relatively low tariffs on imports, no efforts to encourage exports, a deficit budget financed by borrowing from the Central Bank when taxes and aid-generated revenues were insufficient, Central Bank financing of commercial bank credit to the private sector, and low interest rates that assured excess demand for credit.[3]

It is impossible to estimate what would have happened if economic aid had been granted in a manner that gave the government incentives for eliminating its own deficit and encouraging private saving. Several considerations point to the conclusion that, at least through 1957 or 1958, even the best of policies with the benefit of hindsight would have made aid the major source of resource for capital formation. First, if one considers the low per capita incomes of the immediate post-war period, it is apparent that appropriate real interest rate policy could not have induced a higher private saving rate as a fraction of income than was observed in the mid-1960s when income levels were higher and interest rate policy altered; in 1965, private savings were 5.4 percent of GNP, and in 1966, they were 8.3 percent. The latter number is surely an upper-bound estimate (given lower per capita incomes) of what private savings rate could have been achieved; if the government had simply balanced its budget, it would have performed well. The conclusion that aid was essential to generate any growth in per capita incomes follows immediately. And, if a 7–8 percent savings rate is the best that Korea could conceivably have done, aid in excess of 5 percent of GNP was probably necessary to insure a minimal rate of growth of per capita income. However, once aid levels were of that magnitude, it is difficult to imagine that there would not be, directly or indirectly, some consequent disincentives for domestic saving, requiring an even higher aid inflow.

Thus, while there was undoubtedly some avoidable reduction in the domestic savings rate as a consequence of the large aid inflows and the policies surrounding them, it seems reasonable to conclude that large aid inflows were essential, in any event, in the immediate post-war period and that, in their absence, per capita income would have stagnated, if not declined. Had domestic policies regarding savings incentives and the government budget been altered, the realized rate of economic growth probably could have been higher than was in fact the case. To conclude that, however, is not to minimize the crucial role of aid in the post-war reconstruction period.

As to the 1960s and early 1970s, the diminishing relative importance of aid has already been noted. Before borrowing is considered, however, one contribution of aid in the 1950s to growth in the 1960s must be noted: Korea emerged from the 1950s virtually debt free and without any debt-servicing obligations. Had aid in the 1950s been in the form of loans, rather than grants, the prospects for growth in the 1960s would have been significantly diminished, or, alternatively, the same volume of commercial borrowing in the 1960s would have provided far smaller net resources for growth. A simple underestimate of the order of magnitude suffices to illustrate the importance of these considerations. Cumulative U.S. aid from 1954 to 1963 (from Tables 18 and 30) was $2,369 million. If a grace period until 1964 had been extended on all grant aid in the 1950s, and a concessionary interest rate of only 3 percent had been charged, the interest obligation in 1964 would have been $71 million, or 60 percent of exports in that year. An interest rate of 3 percent is probably too low and, had there been interest obligations accruing in earlier years, the debt would have been bigger. But one need not attempt to refine the estimate, for it is difficult to imagine Korea having been sufficiently credit-worthy to enter international capital markets as she did in the late 1960s if a debt of even $2.4 billion had already been incurred. Counterfactual historical experiments are always troublesome, but the importance of Korea's debt-free status in the 1960s should not be underestimated as a contributor to growth during the decade after 1965.

This impression is confirmed by the figures on net borrowing and their contribution to foreign saving during the late 1960s and early 1970s. As Table 52 shows, net borrowing exceeded net transfers starting in 1968, and, by 1970, it was almost three times as large.[4] Frank, Kim, and Westphal attempted to estimate the contribution of foreign savings to growth during the latter part of the 1960s. Two different techniques of estimation were used. First, they noted that foreign savings had fluctuated at around 10 percent of GNP over the 1960–1970 decade. Taking the gross capital-output ratio of 2.5 which prevailed over that period, they concluded that about 4 percentage points of Korea's growth rate could be attributed to foreign savings. As a check on this rough calculation, they estimated an ordinary least squares estimate of real non-agricultural GNP on previous year's non-agricultural GNP and previous year's real investment. The results are in Table 53. The estimated increment in GNP over what would have been realized in the absence of foreign

TABLE 53 Growth Rate and Foreign Savings
             (billions of won)

Actual 1971 GNP	3,151
Estimated 1971 GNP without foreign savings, 1966 to 1970	2,760
Estimated 1971 GNP without foreign borrowing, 1966 to 1970	2,925
Estimated 1971 GNP without foreign commercial borrowing, 1966 to 1970	3,023

Source: Frank, Kim, and Westphal, p. 107.

savings was about 12 percent. When they applied their regression to the 1960s as a whole, the estimate was that about one-third of GNP could be attributed to foreign saving. This econometric result implied a rising incremental capital-output ratio. For this reason, foreign savings in earlier years made a larger contribution to GNP than foreign savings in later years. Consequently, the relative contribution of aid, even in the 1960s, appears greater than simple comparison of transfers and borrowing indicates, since aid was concentrated in the earlier part of the decade.[5]

As the data in Table 52 indicate, domestic savings were rising as a fraction of total savings after 1964. For this reason alone, foreign savings would have been less important in later than in earlier years, especially since the ratio of investment to GNP was also increasing rapidly. Nonetheless, as is evident, the domestic savings rate, even in 1974 and 1975, was still only about 15 percent. Foreign savings continued, therefore, to make an important contribution at the margin to gross capital formation.

Overall, then, aid, virtually the only source of foreign savings in the 1950s, was essential during that decade if there was to be any significant growth in per capita real GNP. By the 1960s, that role was diminishing, although it was still as large as in most other aid-receiving countries. By the late 1960s, commercial borrowing was replacing aid as the chief form of foreign savings, while the domestic savings rate was rising sharply. Whether commercial borrowing could have contributed anywhere near what it in fact did, had aid earlier been in the form of loans rather than grants, is extremely doubtful. Whereas the aid of the 1950s constituted the bulk of available resources for capital formation, the foreign savings of the late 1960s and early 1970s were really a supplement to domestic savings, but one that was critical to permit the high rate of growth that Korea actually enjoyed in those years.

CONTRIBUTION OF EXPORTS TO GROWTH

Whereas it is relatively straightforward to estimate the fraction foreign resources constituted of total saving, and thus infer with some quantitative precision their relative importance in the growth process, any attempt to estimate the contribution of export growth to overall growth is extremely difficult and conjectural. One can, of course, do a straightforward "accounting" of the increment of GNP, and attention will turn to that below. Such an accounting, however, misses a number of intangibles which may be of great importance.

To turn to some of the more obvious considerations, there is the already-mentioned fact that the export-promotion strategy

resulted in a vastly improved resource allocation. This improvement certainly increased the growth rate, but that contribution
cannot be measured by growth accounting. Second, there is the
fact that the foreign savings, which originated through commercial borrowing in the late 1960s and early 1970s, would not
have been available had export earnings not been growing. To a
certain extent, therefore, export growth must be credited with
providing an environment within which Korea was a commercially credit-worthy borrower. Third, there are also such
intangibles as the contact that exporters had with the international market, which may have enabled them to achieve more
rapid increases in productivity and quality control than would
have been possible in the absence of the export orientation.
Fourth, it is difficult, if not impossible, to assess the importance
of the competition Korean exporters faced in the international
market. It is likely that domestic gains in productivity were
more rapid than they would have been in the absence of the
export-oriented strategy. Indeed, the export thrust of policy was
sufficiently pervasive that it would be difficult to find any
aspect of the Korean economy whose performance was not
affected by it.[6]

Along a different line, it is apparent that the dictates of an
export-promotion strategy placed certain constraints on economic policy. The low mean and variance in effective protection
and effective subsidy rates is one such example: the need to
maintain a realistic exchange rate undoubtedly constrained
policy-makers.

It is impossible directly to measure these, and other, intangible effects. What is evident is that the 1950s witnessed at best
a 5 percent rate of growth of real GNP, and even that rate
dropped off sharply once easy opportunities for import substitution had been exhausted and after aid flows leveled off. In
the 1960s rapid growth of exports was accompanied by very
rapid growth of GNP. To be sure, there were numerous pre-
conditions, such as an educated productive labor force, required
to enable the successful growth performance. Nonetheless, it is

noteworthy that rapid growth was occurring against the back-drop of diminishing levels of foreign aid: export growth had to provide foreign-exchange earnings not only for additional imports, but to replace those which had previously been financed by foreign aid.[7]

To turn, then, to the "accounting" measures, the Westphal-Kim estimates of the contribution of exports and import substitutes to growth continue only through 1968. Their findings for the period through 1966 were reported in Chapter 3. For 1966 to 1968, they estimated the percentage contributions as given in Table 54.[8] While these estimates confirm other

TABLE 54 Estimated Percentage Contribution of Exports
& Import Substitutes, 1966–1968

	Total Contribution	Direct Contribution To Manufacturing Growth
Export Expansion	21.3	13.0
Import Substitution	-6.6	-0.1

impressions as to the relative importance of import substitution and export expansion in contributing to growth, they also point to the fact that, for the entire period 1955 to 1968, Westphal-Kim find that domestic demand expansion contributed virtually four-fifths of the total growth in output.[9]

For later years, two sets of estimates are available. The first was made by Wontack Hong. Using national accounts data at 1970 prices, he simply took the ratio of the increment in various magnitudes relative to the increment in GNP. His results, updated to 1975, are presented in Table 55. The first column links the increment in exports to the increment in GNP. As can be seen, that ratio rose from its levels of the mid-1960s to a high of 75 percent in 1972, and remained at very high levels except for 1974. When attention turns to net exports (see Table 36), the orders of magnitude are smaller, naturally, although the same time trend is evident. The fact that net exports in 1974 showed an increase over 1973, while gross exports (in 1970 prices) declined, reflects the fact that, in 1974 (and again in 1975), the ratio of net exports to gross exports increased.

TABLE 55 Direct Contribution of Export Expansion
to GNP Growth, 1963–1975
(% contribution to increase in GNP)

	Increase in Exports	Increase in Net Exports	Increase in Manufacturing	Increase in Value Added of Manufactured Exports
1963	11	—	23	—
1964	11	11	8	3
1965	21	17	41	7
1966	15	-3	20	5
1967	16	7	42	5
1968	17	7	36	5
1969	16	9	26	4
1970	30	16	46	8
1971	24	12	42	8
1972	75	55	53	27
1973	65	29	47	17
1974	-8	6	57	n.a.
1975	54	41	49	n.a.

Source: Wontack Hong, Factor Supply, Table 6.4.

Note: Hong based his computations on the 1970 dollar value of the aggregation. Estimates for 1974 and 1975 are based on 1970 won values from the BOK, Economic Statistics Yearbook, 1976. Data for 1975 are preliminary.

If the data on net exports can be relied upon as an indicator of the trend in the contribution of export expansion to GNP, these numbers can be linked with the Westphal-Kim estimates for 1966 to 1968. From Table 55, it is apparent that the increase in net exports averaged 4 percent of GNP for 1966 to 1968. For that time span, Westphal-Kim show a total contribution, direct plus indirect, of 21 percent. If the same ratio held for 1970 and 1971, it would imply a total contribution, direct plus indirect, of net exports to growth of about 60 percent. For 1972 and 1973, the numbers would be even higher.

The second estimate is from Şuk Tai Suh. His estimates pertain to the direct contribution of exports only. They are reproduced in Table 56. As can be seen, Suh's data also show an increasingly large contribution of exports to growth in the early 1970s—23.6 percent of growth for the 1970–1973 period. For light manufacturing, the contribution was even greater—in excess of 40 percent. Compared with data on the contribution of trade to growth for other countries, these figures are very large.

TABLE 56 Suh's Estimates of the Direct Contribution of Exports, Import Substitution, and Domestic Demand to Growth
(% of total growth)

	1960–1963	1963–1966	1966–1968	1968–1970	1970–1973
Total Growth					
Import Substitution	-2.0	1.7	2.8	0.6	-5.2
Domestic Demand	99.0	89.0	87.8	91.9	81.6
Exports	3.0	9.3	9.4	7.5	23.6
Light Manufacturing					
Import Substitution	6.8	-2.6	-3.3	7.1	-3.9
Domestic Demand	88.7	84.6	84.8	73.9	63.0
Exports	4.5	17.9	18.4	19.0	40.9

Source: Suk Tai Suh, "Import Substitution and Economic Development in Korea," KDI Working Paper 7519, (1975), Table 5-17.

All of these pieces of information point to the importance of export growth as a source of GNP growth. One further question remains: Can anything be said about the extent to which the rate of export growth was, in any sense, optimal? There is already the microeconomic evidence presented in Chapter 5, which indicates that it was fairly close to being so. An alternative approach, using simulation techniques, was employed by

Frank, Kim, and Westphal. They examined, not export levels per se, but rather the extent to which the exchange rate and other commercial policy variables were optimal. In doing this, they first estimated econometrically the behavioral determinants of exports, GNP, and other macroeconomic variables. They then used those estimated equations to simulate changes in policy variables, such as the exchange rate, the height of tariffs, and so on, to ascertain the effects of those alterations upon real GNP.[10]

Their main findings centered around the possibility that, had the Korean government had a slightly higher level of tariffs and lower export subsidies, government savings might have increased, thereby enabling more rapid growth. This finding supports the conclusion reached above with respect to the importance of foreign savings in permitting rapid growth in Korea, since it points to savings as a critical constraint on the rate of growth. With respect to other aspects of commercial policy, Frank, Kim, and Westphal concluded:

> The experiments also support the view that the 1965 exchange rate was an equilibrium rate in the sense that all subsidies and tariffs could have been eliminated and the same historical growth still achieved . . . Our experiments show that the optimal "pure" exchange rate is slightly higher than the actual (about 102 percent of the historical) and is combined with more expansionary monetary and fiscal policies. If subsidies and taxes on exports and imports are combined with exchange rate policy, the optimal rate is about equal to the historical rate. The optimal rate should be combined, however, with higher import duties (or fewer exemptions) and roughly similar subsidies.[11]

This finding tends to support the conclusions emerging from analysis of microeconomic data: there is little evidence of any significant inefficiency in the Korean push to promote exports. On the contrary, it would appear that exports contributed significantly to the growth rate of GNP, but that any higher growth of exports would have been at the expense of GNP growth. While there are always ways in which one can, with hindsight, find improvements that might have been made in policy variables, it is difficult to point with any certainty to any major changes in Korean economic policies with respect to the trade-and-payments regime that could have significantly increased the rate of growth of real output.

RELATIONSHIP OF EXPORTS
AND EMPLOYMENT GROWTH

One of the remarkable features about the altered performance of the Korean economy after the export-promotion drive started compared to its earlier behavior is the rapid growth of non-agricultural employment opportunities. Table 57 gives the basic data on growth of the population and employment. It is estimated that, in the early 1960s, unemployment reached 8.3 percent of the labor force with about 57 percent of the population engaged in farming.[12] That rate, therefore, was equivalent to about 17 percent of the non-farm labor force seeking jobs. As Table 57 shows, non-farm employment rose rapidly, more than doubling between 1964 and 1975. Whereas farm employment exceeded non-farm employment almost 50 percent in 1964, non-farm employment was about 30 percent greater than farm employment in 1975.

Moreover, the unemployment rate fell dramatically, reaching a low of 4 percent of the labor force in 1973. Since by that time the population was more than 50 percent non-farm, that rate was equivalent to about 7.3 percent of the non-farm labor force.

In view of the fact that Korea's success with increasing employment opportunities contrasts sharply to the experience of the majority of developing countries where slow employment growth is a major problem, it is natural to investigate the relationship between employment growth and the export-promotion strategy of the 1960s. Of course, to the extent that rapid overall growth resulted from export growth, and employment

TABLE 57 Labor Force Data, 1964–1975
(1,000s of persons)

	Population 14 and over	Agricultural Employment	Manufacturing Employment	Total Non-Farm Employment	Total Employment	Unemployed Number	Unemployed % of Labor Force
1964	15,052	4,655	637	3,144	7,799	650	7.7
1965	15,937	4,603	772	3,603	8,206	653	7.4
1966	16,367	4,695	833	3,728	8,423	648	7.1
1967	16,764	4,598	1,021	4,119	8,717	578	6.2
1968	17,166	4,582	1,176	4,573	9,155	492	5.1
1969	17,639	4,687	1,232	4,727	9,414	474	4.8
1970	18,253	4,826	1,284	4,919	9,745	454	4.5
1971	18,984	4,758	1,336	5,308	10,066	476	4.5
1972	19,724	5,110	1,445	5,449	10,559	499	4.5
1973	20,438	5,260	1,774	5,879	11,139	461	4.0
1974	21,148	5,304	2,012	6,282	11,586	494	4.1
1975	21,833	5,123	2,205	6,707	11,830	510	4.1

Source: BOK, *Economic Statistics Yearbook, 1976*, Table 135.

opportunities grew rapidly because the economy grew rapidly, one can in any event attribute employment growth to the export-promotion strategy. Nonetheless, it is worthwhile to inquire how much of the increment in non-farm, and especially manufacturing, employment is directly and indirectly attributable to the demands generated by the export sector.

It has already been seen that manufacturing exports became increasingly labor-intensive over the period 1960 to 1968. This was the result of two interdependent factors: 1) there was a shift in the commodity composition of exports toward labor-intensive commodities; and 2) the real wage appears to have remained relatively constant during the early years of rapid growth of manufacturing. [13] This enabled the upward shift in the demand for labor to be reflected in increasing employment opportunities, rather than in rising real wages for those already employed. The more rapid rate of increase in real wages after 1966 was primarily a consequence of market forces, reflecting the fact that unemployment had diminished and wages had to rise to attract additional workers. Once wages began rising, it is not surprising that the incremental capital-output ratio, in export industries as in others, began rising, as reflected in the capital-labor ratios discussed in Chapter 5.

Against this background, it is of interest to examine estimates of the contribution of exports to employment growth. Two such sets of estimates have been made. On one hand, Watanabe [14] relied upon sample-survey data in order to attempt to differentiate between exporting and domestic-market activities within industries. His estimates are for 1969 only. On the other hand, Cole and Westphal [15] relied upon input-output data to derive estimates of the direct and indirect employment generated by exports, and to provide comparable estimates for 1960, 1963, 1966, and 1970.

Each of these methods has advantages and drawbacks. The sample-survey method is especially weak when it comes to estimating indirect effects. The input-output method is distinctly superior in estimating indirect employment effects of exports, but is less satisfactory for distinguishing between labor employed within an industry to produce for the domestic market, and labor employed to produce export commodities.[16] Despite their differences, the two methods of estimation yield very similar figures for total employment attributable to manufacturing for the one year that both estimates are available.[17] Because of this, and the fact that the Cole-Westphal estimates cover more years, only their results are considered here. Table 58 gives their basic results.

TABLE 58 Estimated Employment Attributable to Exports, 1960, 1963, 1966, 1970

(1,000s of workers and %)

	1960	1963	1966	1970
Primary Sectors				
Direct Employment in Exports	128	71	75	108
Total Employment Due to All Exports	214	181	237	279
% of Primary Employment	3.4	2.7	4.4	5.0
Manufacturing				
Direct Employment in Exports	12	23	113	225
Total Employment in Exports	26	46	172	348
% of Manufacturing Employment	5.0	6.4	16.5	22.5
All Sectors				
Direct Employment in Exports	183	134	274	475
Total Employment in Exports	302	290	585	941
% of Total Employment	3.7	3.3	6.7	8.9

Source: Cole and Westphal, "The Contribution of Exports", p. 94.

Note: Cole and Westphal provide two sets of estimates for 1966 and 1970, but only one for 1960 and 1963. To maintain comparability, only the set available for all years is reproduced here.

The most significant impact of exports on employment is clearly within the manufacturing sector, where Cole and Westphal estimate that the number of jobs directly attributable to manufactured exports rose from 12,000 in 1960 to 225,000 in

1970, while total employment due to exports rose from 26,000 to 348,000. This represented an increase from 5 percent of total manufacturing employment in 1960 to 22.5 percent in 1970. Presumably, this figure rose even higher with the export boom of the early 1970s.

For the economy as a whole, approximately 9 percent of all jobs were attributable to exports in 1970, compared to less than 4 percent in 1960. That 9 percent of all jobs are export-related may not seem startling at first glance, but it is much more impressive when it is recalled that the increase in export-related employment was greater than the decrease in unemployment in the period in question.

There are, of course, difficulties in attributing a particular fraction of employment growth to export growth. Had exports not expanded as rapidly, it is likely that job opportunities would have been created elsewhere in the economy. Moreover, employment growth itself was a function of the behavior of the real wage and not simply of exports. Indeed, had the real wage risen rapidly with the beginning of the export boom, the boom itself might have been thwarted, since the Korean comparative advantage in labor-intensive commodities might not have been reflected in wage costs.

Nonetheless, it seems clear that rapid export growth contributed significantly to expanding employment opportunities throughout the years of the export-promotion strategy. In addition to that direct effect, there was the contribution provided by foreign savings, in the form of enabling more investment and more jobs, and the contribution of more rapid growth, part of which can be attributed to export growth.

CONCLUSIONS

Techniques are not available for estimating the "true" contribution of the export-oriented growth strategy to Korea's rapid growth during the 1960s and early 1970s. A "true" estimate would entail the specification of an alternative growth path over those years, and an estimate of what would have happened under an alternative strategy.

In a sense, though, pointing to what would be required for an accurate assessment of the role of the trade-and-payments regime to the Korean economy highlights an important aspect of that role: the interrelationships between the export orientation of the economy and virtually every other economic variable are so close and so complex that the question is largely unanswerable. Korea adopted an export-promotion strategy in 1960. Thereafter, her growth rate increased markedly. Other reforms were undertaken which were probably essential to the continued success of the export drive. Certain necessary conditions in the domestic economy were also met, including the availability of an industrious and literate labor force and the willingness of the government to allow market forces to determine the wage. Initial success led to more rapid export growth, and more rapid growth of GNP. That, in turn, led to further rapid export growth. The commitment of the government to the export strategy was so complete that virtually all policies were scrutinized and considered in light of their implications for the export drive.

Description of what happened, however, does not necessarily imply causation. Yet attempts to estimate causation by quantifying the macroeconomic contribution of the trade-and-payments regime, aid, and capital flows leave the inescapable impression that some important attributes of each are not captured. Even so, foreign savings and the export drive have each been crucially important by these measures.

The role of foreign savings in permitting larger gross capital formation than would otherwise have occurred has been of great importance throughout the thirty years of Korean modernization. While foreign savings, in the form of aid in the 1950s, were the predominant source of resources for investment, their role continued significant in permitting a very high growth rate right until the end of the thirty-year period of modernization.

Obviously, in the 1970s Korea could have sustained a positive rate of growth in the absence of a capital inflow, but foreign savings permitted a much higher growth rate than would otherwise have been possible. That role was different from the role in the 1950s, when aid was virtually the only means of obtaining imports and providing resources for capital formation.

Export growth, of course, did not begin rapidly until after 1960. Thereafter, it was *the* most salient characteristic of the Korean economy. While its influence was pervasive in many intangible ways, quantitative estimates suggest that between 4 and 8 percentage points of the growth rate were attributable to export growth, at least until the early 1970s. In those years, the contribution may have been even greater. Certainly the modernization of Korea would have proceeded much more slowly if exports had grown more slowly. Finally, rapid export growth was a significant factor in permitting the rapid growth of employment opportunities.

Whether these quantifiable contributions of trade, aid, and capital flows to Korea's modernization are necessarily their most significant contributions is a matter for debate. But, by any standard, trade, commercial policy, aid, and capital flows have been integrally linked to Korea's fortunes throughout the modernization period.

Appendix A
Definition of Exchange-Rate Terms

It usually happens, especially in contexts of exchange control, that the official parity of a country bears little relationship to the actual receipts of an exporter or costs of an importer per unit of foreign currency. Moreover, in the context of a rate of inflation significantly different from that in the rest of the world, the economically meaningful measures of exchange rates in units of local currency need deflation to render them comparable over time. For these reasons, it is useful to distinguish among the exchange-rate concepts given below.[1]

1. *Nominal exchange rate*: The official parity for a transaction. For countries maintaining a single exchange rate registered with the International Monetary Fund, the nominal exchange rate is the registered rate.

2. *Effective exchange rate (EER)*: The number of units of local currency actually paid or received for a one-dollar international transaction. Surcharges, tariffs, the implicit interest foregone on guarantee deposits, and any other charges against purchases of goods and services abroad are included, as are rebates, the value of import replenishment rights, and other incentives to earn foreign exchange for sales of goods and services abroad.

3. *Price-level-deflated (PLD) nominal exchange rates*: The nominal exchange rate deflated in relation to some base period by the price level index of the country.

4. *Price-level-deflated EER (PLD-EER)*: The EER deflated by the price level index of the country.

5. *Purchasing-power-parity adjusted exchange rates*: The relevant (nominal or effective) exchange rate multiplied by the ratio of the foreign price level to the domestic price level.

Appendix B
Important Dates for Trade and Aid
in Korea's Modernization

September 1945: U.S. Military Government installed in Korea.
1947: First stage of land reform.
June 1947: Chosŏn Exchange Bank established to facilitate private foreign trade.
August 1948: Transfer of authority from U.S. Military Government to the Republic of Korea.
January 1949: U.S. Economic Cooperation Administration begins functions.
June 1950: Beginning of war between North and South Korea.
July 1953: Korean War Armistice signed.
December 1953: Wŏn officially devalued from 6 to 18 per dollar.
August 1955: Wŏn officially devalued to 50 per dollar.
1958–1959: Stabilization program cuts growth of real output.
April 1960: President Rhee resigns.
January 1961: Devaluation of wŏn from 65 to 100 per U.S. dollar.
February 1961: Devaluation of wŏn from 100 to 130 per U.S. dollar.
May 1961: Military coup from which General Park emerges as head of ruling junta.
June 1961: Unification of multiple exchange-rate system.
January 1963: Return to multiple exchange rates.
October 1963: Elections after which General Park is elected President by National Assembly.
May 1964: Devaluation from 130 to 257 wŏn to the dollar; fluctuating exchange rates.
March 1965: Reunification of multiple exchange rates.
1965: Normalization of relations with Japan.
September 1965: Interest rate reform.

July 1967: Reform of the import control system from positive-list to negative-list system.

1967: Tariff reform.

June 1971: Devaluation from 326 to 370 wŏn to the dollar; exchange rate pegged.

June 1972: Exchange rate pegged at 400 per dollar after controlled upward floating.

Fall 1973: Oil price increases.

December 1974: Wŏn devalued to 484 per dollar.

July 1975: Shift from duty exemption to drawback system for exports.

Notes

Preface

1. Charles R. Frank, Kwang Suk Kim, and Larry E. Westphal, *Foreign Trade Regimes and Economic Development: South Korea* (New York, 1975).

2. Larry E. Westphal and Kwang Suk Kim, "Industrial Policy and Development in Korea," (World Bank Staff Working Paper No. 263, August 1977).

3. See Wontack Hong, "Trade, Distortions and Employment Growth in Korea," (KDI, mimeo, 1977), and Statistical Appendix thereto; and Suk Tai Suh, "Import Substitution and Economic Development in Korea," (KDI, mimeo, 1975), and "Foreign Assistance in Modernization of Korea," (KDI, mimeo, 1976).

ONE   *The 1945–1953 Period*

1. The basic source of these data is Chōsen Sōtokufu, *Chōsen Sōtokufu tōkei nenpō*, and H. Ouchi, ed., *Nihon keizai tokeishū*, (Tokyo, 1958). They are presented in Wontack Hong, "Trade, Distortions and Employment," Statistical Appendix, Tables B.1–B.16, (Seoul, mimeo, 1977).

In 1936, 52% of gross "commodity production" originated in agriculture, 5% in forestry, 7% in fisheries, 5% in mining, and 31% in manufacturing. According to Hong's estimates, in value-added terms, about 64% of production originated in agriculture and about 15% in manufacturing in 1936. Non-commodity sector output is omitted from the computation. If non-commodity sector constituted 30% of GNP, agriculture would have accounted for about 45% of national income and manufacturing for about 11%. Hong, "Trade, Distortions and Employment," Chap. 2 and Tables B.1–B.2.

2. Ibid. Manufactures exports from Korea included some re-exports of Japanese-made products which amounted to around 5% of total manufactures exports.

3. Ibid., Table B.8.

4. Ibid., Table B.5.

5. The term "Japan proper" refers to Japan itself, Taiwan, and her other colonial islands in the Pacific.

6. Wontack Hong, "Trade, Distortions and Employment." In the years 1901 to 1905, exports to Japan proper were about 80% of total Korean exports, while about two-thirds of all imports originated from Japan. The fraction of trade with Japan proper was thus high prior to colonial rule, although it rose still further during that period.

7. Korean yen became Korean wŏn in 1945. A currency reform on February 14, 1953, exchanged 100 wŏn for one hwan. A later reform, in 1961, exchanged one new wŏn for ten hwan. Throughout this study, units will always be new wŏn.

8. Based on Hong's approximation of per capita GNP over the period 1910 to 1940. See Hong, "Trade, Distortions and Employment," Tables B.1.–B.3.

9. Ibid., Table B.11.

10. BOK Research Department, *Price Statistics Summary* (1964), p. 23. The index, with 1960 = 1, stood at 0.120 in December 1939 and 0.188 in June 1945. This would imply that inflation during World War II was relatively modest. Indeed, all accounts suggest that damage during World War II was not large relative to that inflicted upon many other countries. On a 1960 = 100 base, the price index stood at 0.076 in August 1945 (which would compare with 0.0018 two months earlier) and 0.331 in December 1946.

11. There does not appear to be a satisfactory commodity breakdown of exports for that period. The data for 1949 indicate that 69% of the wŏn value of exports consisted of food products and 13% inedible crude materials. The proportions in 1950 are not significantly different. See BOK, *Economic Statistics Yearbook, 1958*, p. 213.

12. The last U.S. occupation forces were withdrawn in 1949, although technical military advisers continued to be supplied to the ROK under ECA auspices.

13. For a description of the incoherence and lack of coordination of American post-war policy toward Korea, see Gregory Henderson, *Korea: The Politics of the Vortex*, (Cambridge, Mass., 1968), Chap. 5. Until the end of 1947, U.S. policy was based on the implicit assumption that

reunification would take place. Thereafter, the debate centered over whether Korea was in any sense vital to U.S. interests. See Henderson, pp. 148 *ff.*

14. Harold Koh, "The Early History of U.S. Economic Assistance to the Republic of Korea, 1955–63," typed, 1975 (p. 2) gives this figure, citing W. D. Reeve, *The Republic of Korea: A Political and Economic Study* (London, 1963), p. 105.

15. Apparently, there had earlier been discussion of a $500 million aid program to start in 1947 that would concentrate on development objectives. This was regarded as part of an overall strategy that would strengthen South Korea and enable the United States to withdraw its military units, and leave South Korea outside the "defense shelf" of the United States in the Pacific. The British withdrawal from Greece and Turkey at that juncture inspired President Truman's "Point Four" proposal. However, "President Truman quietly informed the Department of State that he could not go before Congress with two large requests; the Korean long-term aid program would have to be dropped. After that time, until the Korean War, 'our support for Korea tottered along in an unimpressive, inadequate, and sporadic fashion.' A decision had been reached: . . . disengagement took place, but the 'gracefulness' was missing: without requisite aid or defense the Korean policy created by Americans was a legless monster from birth." Henderson, p. 150.

16. E. A. G. Johnson, *American Imperialism in the Image of Peer Gynt: Memoirs of a Professor-Bureaucrat* (Minneapolis, 1971), p. 178. Johnson was Director of the Korean ECA program after earlier serving in various roles in the interim government, including Minister of Commerce.

17. U.S. Department of State and the Economic Cooperation Administration. "ECA Recovery Program for 1950" (mimeo, Washington, June 1949), p. 4.

18. General Wedemeyer, among others, strongly expressed this view. See Koh, p. 3.

19. When responsibility for administration of U.S. aid was transferred to ECA, the Republic of Korea and the American government signed the ROK-U.S. Agreement on Aid. This agreement was similar to the ones used under the Marshall Plan and in effect stipulated that the two governments would achieve a consensus on desirable monetary, fiscal, and balance-of-payments behavior. This issue already became a bone of contention in 1949. It is discussed below.

20. Sung Hwan Ban, *Growth Rates of Korean Agriculture 1918–1971*, (KDI, 1974), pp. 245–250.

21. Henderson, p. 156. Earlier rentals were generally between 50 and 90% of output, although owners supplied purchased inputs and maintenance.

These expenses are estimated to have amounted to about 20% of gross output, so that net rentals were probably in the range of 40–70% of output.

22. For a number of reasons, the occupation forces became identified with the earlier colonial Japanese rulers. A very early statement that USAMGIK would distribute the land implied that it would not seek compensation. USAMGIK consistently underestimated the technical problems involved in divestiture and even, at one time, contemplated retaining title to the land to turn over to the Korean government when occupation ended. This led to considerable misinterpretation of American intentions, which in turn provided the pressure for immediate distribution. For further details, see Koh, Chapter II.

23. Henderson, p. 156. 1 chŏngbo = 2.45 acres.

24. Clyde Mitchell, "Land Reform in Asia, a Case Study," (National Planning Association Pamphlet 78, 1952), pp. 19–20.

25. See *Rural Development* in this series for an in-depth analysis of land reform.

26. Koh, p. 27.

27. David C. Cole and Princeton N. Lyman, *Korean Development, The Interplay of Politics and Economics*, (Cambridge, Mass., 1971), p. 21.

28. Henderson, pp. 156–157.

29. Cole and Lyman, pp. 21–22.

30. Koh, p. 66.

31. For greater detail, see in this series Noel F. McGinn, et al. *Education and Development in Korea.*

32. See in this series Kwang Suk Kim and Michael Roemer, *Growth and Structural Transformation* for a comparison of the real per capita consumption levels of the 1930s with those of 1910 and the period before Japanese rule began.

33. It should not be confused with the United Nations Commission on the Unification and Reconstruction of Korea. This Commission, as its name implies, was charged with the mission of formulating plans for development in the event of reunification of the country.

34. For a detailed history of UNKRA and its attempts to function during the war, see Gene M. Lyons, *Military Policy and Economic Aid: The Korean Case, 1950–1953* (Columbus, Ohio, 1961).

35. John P. Lewis, "Reconstruction and Development in South Korea," (National Planning Association, December 1955), Pamplet 94, p. 36.

36. The issue could be turned around, and *all* expenditures of an ally to assist in defense of territory could be classed as aid. This demonstrates the futility of attempting a definition.

37. It can be argued that the burden of the war effort was not otherwise fully shared, but that issue takes us far afield here and relates to

the more general and unanswerable question as to how aid should be defined.

38. See Lewis, pp. 39 *ff.*, for more details.

39. Donald G. Tewksbury, *Source Materials on Korean Politics and Ideologies*, (New York, 1950), pp. 145–46. One of ECA's activities in 1949–1950 was to sponsor the visits of Arthur Bloomfield and John P. Jensen, then both of the Federal Reserve Bank of New York, to assist in the formulation of banking legislation and the foundation of the Bank of Korea. See Koh, Chapter III, and Arthur I. Bloomfield and John P. Jensen, "Banking Reform in South Korea" (New York, March 1951).

40. Devaluation would also have increased the value of counterpart funds, which provided yet another reason for reluctance to devalue: more counterpart funds meant that the United States had to approve more Korean government expenditures.

41. Frank, Kim, and Westphal, p. 15.

42. See Appendix A for the definition of effective exchange rate.

43. Strictly speaking, the appropriate index ought to be a "purchasing-power-parity" price level deflated *effective* exchange rate. During the period under review, there was an increase in the world price level and also that of Korea's major trading partners. However, it was small relative to fluctuations in the real exchange rate in Korea. The fact that nominal, and not effective, exchange rates are used in Table 8 is probably the more serious omission.

44. See note 42.

45. Until then, the wŏn redemption rate was negotiated separately for each advance. Conditions deteriorated so far in the fall of 1952 that the South Korean government suspended advances on December 15, advising the UNC to redeem its accumulated advances and to buy wŏn from the Bank of Korea. In return, the flow of petroleum products for civilian use was halted. See Frank, Kim, and Westphal, pp. 28–29, 41.

46. Tungsten, which was the chief mineral export, was exported only under government monopoly.

47. See Frank, Kim, and Westphal for additional details, p. 26 *ff.*

48. Ibid., p. 34.

49. It is hard to estimate how important customs duties were as a source of revenue. I was able to locate data only for 1953. In that year, customs duties receipts were 351 million wŏn, while total internal tax receipts were 1,745 million wŏn. However, BOK borrowings and bonds issued were 2,723 million wŏn. Data are from Wontack Hong, "Trade, Distortions and Employment," Table B.24.

50. It will be recalled that, even earlier, "trust shipping" had provided such a link under the Bank of Korea.

TWO   *Trade and Aid, 1953 to 1960*

1. Analysis of the efficiency of resource allocation under import substitution is provided in Chapter 5.

2. See Table 8 above.

3. See Appendix A for definitions of the various exchange rate measures and concepts.

4. See Appendix A for definitions.

5. Capital flows, other than aid, were virtually nonexistent.

6. See pp. 56, 161ff.

7. In any country with quantitative restrictions, the government allocates foreign exchange among competing import claims. It is not obvious why there were additional difficulties resulting from the fact that foreign exchange was received initially by the government. However, Frank, Kim, and Westphal, (p. 29), comment thus: "The large inflow of U.S. grant aid, United Nations Korean Reconstruction Agency (UNKRA) assistance, and government receipts of foreign exchange from United Nations Command (UNC) sources created difficulties in allocating foreign exchange to various industrial sectors that lasted for some time after the war." Kwang Suk Kim, in correspondence, has suggested that U.S. fiscal year requirements and regulations surrounding the use of aid funds were the chief source of difficulty.

8. Sources are: Wontack Hong, "Trade, Distortions and Employment" Table B.24 for customs duties; IMF, *International Financial Statistics,* May 1976, for won value of imports.

9. Frank, Kim, and Westphal, pp. 29–34.

10. Ibid., p. 34.

11. There were also a few exceptions wherein commodities were imported at *less* than the official rate. Such was the case with fertilizer which, until February 1956, was 25 wŏn per dollar, although parity had earlier altered to 50 wŏn per dollar. See IMF, *Annual Report on Exchange Restrictions, 1956,* p. 216.

12. See pp. 165–166 for an estimate of the order of magnitude of premiums.

13. IMF, *Annual Report on Exchange Restrictions, 1956,* pp. 219–220.

14. IMF, *Annual Report on Exchange Restrictions, 1961,* p. 228.

15. Frank, Kim, and Westphal, p. 234.

16. U.N., *Yearbook of International Trade Statistics, 1955, 1956,* and *1957,* Korea Tables, Table 3, pp. 457, 357, and 357 respectively.

17. The food and beverages, and textile sectors' exports increased sharply after 1957 and 1958 respectively. However, textile exports in

1959 were still below the level of any year between 1953 and 1957.

18. The data in Table 13 were derived from a 43-sector classification. Aggregation to the level given in Table 13 was done primarily to avoid having entire rows of zeros. Reproduction of the entire 43-sector classification would only reinforce the impression of erratic behavior on the part of individual exports.

19. Frank, Kim, and Westphal, p. 92.

20. Ibid., pp. 96–97.

21. In part, this may reflect the fact that some supplies of consumer goods were imported by the military and not included in commercial imports. Supplies diverted from the PXs may also have increased domestic availability. Both these factors would imply a downward bias in Suk Tai Suh's estimate of the ratio of imports to domestic consumption.

22. Kwang Suk Kim, "Outward-Looking Industrialization Strategy: The Case of Korea," in Hong and Krueger, *Trade and Development,* p. 20.

23. Suk Tai Suh, *Import Substitution,* Tables 5-3-1, 5-3-2, and 5-3-3.

24. Primary industry has a higher value-added content than manufacturing, so that these figures probably overstate the increase in the relative importance of manufacturing.

25. The Korean statistics provide a breakdown of imports into "commercial," "official aid," "foreign loans," and "other." On the basis of those data, aid imports were 55, 61, 68, 83, and 85% of total imports for the years 1953 to 1957.

26. There are also problems, as demonstrated in Suk Tai Suh's appendix, with the appropriate measure of aid. There are fiscal year and calendar figures, each on an obligation basis, a disbursement basis, and a delivery basis. Differences in timing can affect the yearly totals, sometimes by substantial magnitudes.

27. See pp. 74–75 for a discussion of the use of counterpart funds.

28. Data for 1953 are not available.

29. The source for this statement is data provided by Suk Tai Suh, *Import Substitution.* The sector for which commodity project support was greatest was Public Utilities, which received more than two-thirds of the total in all years except 1954. The Fertilizer sector was the largest recipient of project supporting assistance for plant.

30. Data are from ibid., Table 5-2.

31. Ibid.

32. It will be recalled that percentages do not add to 100 due to the "unclassfiable" category. The source is the same as Table 20.

33. Moreover, the fact that data on EERs are unreliable makes the entire effort suspect. For even in cases where the commodity is homogeneous,

a comparison of the ratio of the domestic price to the imported price at the official exchange rate tells little.

34. See also Chapter 5, where the results of an effort to estimate via simulation are reported.

35. When "freed resources" are then allocated to another purpose, there is a resource allocation effect. In that instance, the effect of the imports is not generally deflationary. When the counterpart funds are accumulated rather than spent, the effect is deflationary. Difficulties arise when counterpart funds are accumulated and then spent at a later date: in the first period, the effect of the imports is deflationary; when counterpart funds are then spent, the offsetting resources are already absorbed, and the effect is not dissimilar to that from printing money.

36. Data are from Cole and Lyman, p. 174. Borrowing constituted the remaining 13% of government receipts. Government expenditures exceeded government revenues by 50%, with much of the excess financed by U.S. military assistance.

37. For a description of the plan, see Lewis, Chapter IV.

38. Ibid., pp. 35–36.

39. Koh, p. 13.

40. Cole and Lyman, pp. 164–165.

41. Ibid., pp. 167–168.

42. Ibid., p. 129. See Chapter 6 for further discussion.

43. Ibid., p. 165.

THREE   *The Transition to an Export-Oriented Economy*

1. See Tables 9 and 10. As discussed in Chapter 2, as of 1960 the premium-exclusive EER for exports was above that for imports, although there is every reason to believe that the system was still biased toward import substitution.

2. Kwang Suk Kim, pp. 25–26, in Hong and Krueger, *Trade and Development.*

3. Insofar as quantitative controls still left sizable premiums on licenses for some import commodities, a variable exchange tax was levied in June 1961 to absorb the premiums. See discussion under quantitative restrictions below.

4. The data in Table 22 reflect annual averages of exchanges rates and are not end-of-year figures.

5. In fact, some vestiges of multiple rates continued, as exchange certificates were sold in the curb market. However, the relative importance of the secondary rates was markedly reduced.

6. The link system was reinstituted in 1963, resulting in a return to multiple exchange rates.

7. To try to absorb the implicit value of restrictions, a monthly survey of prices was taken and commodities were reclassified, depending on the results. On the basis of the 1964 results, the number of items was increased to 2,700. See Frank, Kim, and Westphal, p. 49.

8. However, the fact is that even in this case the tariff exemption covered more than the intermediate goods necessary for production of exportables. The exporters were therefore able to profit by this means when they sold on the domestic market.

9. Some "incentives" to export were of a different form. Firms failing to meet their expected performance in exporting experienced a number of difficulties in other dealings with the government. The value of the the intangible "government approval" incentive was, and continues to be, considerable.

10. Although no breakdown is available of the relative importance of private and public enterprises as a source of exports, there is every indication that the public enterprises' share of exports, even when adjusted for the sectoral composition of output, was far less than their share of output. See Leroy Jones, *Public Enterprises and Economic Development: The Case of Korea* (Seoul, 1975), pp. 114 ff.

11. See Chapter 5 for a further discussion of the effects of the wastage allowance provisions.

12. Frank, Kim, and Westphal, p. 46.

13. While it is impossible to provide any quantitative estimate of the significance of targets, they surely played a considerable role. At the time the modernization study was under way, targets for exports during the period 1976–1981 were being debated. It was generally thought that MCI officials wanted relatively low targets since higher targets would entail "more work" for them.

14. See Frank, Kim, and Westphal, Table 5–8.

15. Data are from Wontack Hong, "Statistical Appendix," p. C.39, where the percentage distribution of commodity exports by destination is given. These percentages were then multiplied by the export totals given in Table 13 and Table 25 to estimate total exports to Japan.

16. Contrasted with most other countries, Korea was able to adjust remarkably well to the oil price increase and worldwide recession of the mid-1970s, as will be seen in Chapter 4. By that time, Korea was already well established in international markets, but her success in adjusting to those events does show that Korean policies are adapted to international conditions and that success might have occurred even in less favorable world market conditions in the 1960s.

17. Crop failure and the consequent increase in food imports also contributed to the increase in imports.

18. Petroleum imports are not to be confused with imports of petroleum products.

19. Suh, *Import Substitution,* Table 5-2. There are also a number of categories for which the import-domestic demand ratio rose. Most notable is transport equipment, for which the percentage of imports rose from a very low level in the late 1950s to 25% by 1965 and even higher proportions in the late 1960s. The fraction of machinery and electrical equipment imported also began rising after 1965, as did fabricated metal products and basic chemicals.

20. Frank, Kim, and Westphal, p. 92.

21. See Chapter 6.

22. Frank, Kim and Westphal, p. 105.

23. Ibid., p. 104.

24. Cole and Lyman, p. 90.

25. Data are derived from Suk Tai Suh, *Import Substitution,* Tables II-6 and II-7.

26. See Kwang Suk Kim and Michael Roemer, *Growth and Structural Transformation,* Studies in the Modernization of the Republic of Korea: 1945–1975 (Cambridge, Mass., 1979) for a full discussion of the financial and monetary reforms of 1964–1965.

27. Even so, some distortions were introduced into the payments regime by virtue of a differential between the domestic and foreign interest rate. See Chapter 4.

FOUR   *Emergence as a Major Exporter, 1966 to 1975*

1. Kwang Suk Kim, "Outward Looking Industrialization Strategy," p. 21.

2. Frank, Kim, and Westphal, Chapter 5.

3. Data are from IMF *International Financial Statistics,* (May 1976). The export target in the Fourth Five-Year Plan was to achieve a 1% share of world trade by 1981.

4. IMF, *Annual Report on Exchange Restrictions, 1970,* p. 297.

5. Ibid., *1971,* p. 256, and *1972,* p. 294.

6. IMF, Ibid., *1973,* p. 295.

7. IMF, Ibid., *1975,* p. 297.

8. Another factor that may have been significant was the increasingly protectionist stance of the United States. In January 1972, the Korean and

American governments signed a five-year agreement, retroactive to 1971, under which Korea would limit the annual growth of synthetic and woolen textile exports to 7.5% and 1% by volume respectively.

9. See pp. 000–000 for an examination of the tariff structure. In addition to the tariff schedules, the government continued to have and to employ its power to administer variable tariffs to absorb the premiums on import licenses.

10. The wastage allowance component of the exemption was extended to the rebate system.

11. Cole and Lyman, pp. 190–191.

12. Larry E. Westphal and Kwang Suk Kim, "Industrial Policy and Development in Korea," (mimeo, 1974), p. 9.

13. The tariff rates upon which import EERs were calculated are taken from actual tariff collections: if the legal rates had been used, the differential incentive in favor of exports would appear to be somewhat smaller.

14. IMF, *Annual Report on Exchange Restrictions, 1968*, p. 254. Frank, Kim, and Westphal (p. 58) give the following data in an attempt to compare the situation before and after the shift:

Number of Automatic Approval Sub-items:

Before July 24, 1967	*After July 25, 1967*
3,760	17,128

15. IMF, *Annual Report on Exchange Restrictions, 1969*, pp. 274–275.

16. Ibid., pp. 274–275.

17. Frank, Kim, and Westphal, pp. 56–57. The attempted tariff reform and the debt-servicing problem are discussed on pp. 140–141, 146–148.

18. Their percentage was even higher in earlier years, reaching 38% in 1973.

19. Westphal and Kim, p. 4-4.

20. See Frank, Kim, and Westphal, p. 82.

21. Frank, Kim, and Westphal (pp. 82–83) provide some additional evidence that tends to confirm this view.

22. A frequently heard assertion is that South Korean sales to Vietnam were an important explanation for South Korea's success in promoting exports. Sales started in 1967 ($15 million) and reached a peak of $64 million in 1971—hardly a major part of the export boom.

23. IMF, *Annual Report on Exchange Restrictions, 1972*, p. 260.

24. Data were taken from BOK, *Economic Statistics Yearbook 1976*, pp. 260–261.

25. In 1976, yet another attempt to reform the tariff system was started. See Suk Tai Suh, "Revision of Tariff Rates and the Introduction of Flexible Tariff System" (KDI, mimeo, July 5, 1976) for more details.

26. Frank, Kim, and Westphal, p. 50.

27. IMF, *Annual Report on Exchange Restrictions, 1974,* p. 268.

28. Hong, "Statistical Appendix," Table B-39.

29. This authority was used in 1974 and 1975. Tariffs were raised on items competing with domestic production, and lowered on raw material imports. The latter was designed to offset part of the impact of shifting from customs exemption to a drawback system. See Suk Tai Suh, *Import Substitution,* for an itemization.

30. See Chapter 5 for estimates of effective rates of protection and the resource allocation effects of tariffs.

31. Development Loan Fund sources from AID were the main exception.

32. The normalization of relations with Japan in 1965 also contributed to increased capital flows. Under the agreements, the Japanese were to provide $300 million in credits to Korea. Of course, Japanese were also eligible to provide equity capital under the same conditions as other foreigners, but no amount was stipulated under the agreement.

33. This naturally had implications for resource allocation, which are discussed in Chapter 5.

34. Data are from Frank, Kim, and Westphal, Table 7-5, p. 116.

35. Parvez Hasan, *Korea, Problems and Issues in a Rapidly Growing Economy* (Baltimore, 1976), p. 251.

36. Although there was discussion of debt-management problems in 1970–1971 and 1974–1975, it would appear that concern was aroused by behavior of the current account and not by debt-service obligations themselves.

37. As seen earlier, the opposite was true of direct investment, where Japanese investors accounted for 66% of all direct investment and American investors accounted for 27%.

38. Frank, Kim, and Westphal, p. 106. This was in addition to the $300 million in commercial credits mentioned earlier.

39. The International Bank for Reconstruction and Development was a relatively unimportant lender until 1968. Even thereafter, its loans were moderate as a fraction of Korea's overall public indebtedness. As of Dec. 31, 1974, public debt outstanding to the World Bank was $492 million, of which $224 million had been disbursed. This represented 8% of the total public debt and 5.5% of disbursed loans. See Hasan, p. 221.

40. Hasan's data on public debt at the end of 1974 indicate that about half was owed to other governments and international institutions and half to private creditors. See pp. 220–221.

FIVE   *The Allocative Efficiency of Trade and Aid*

1. It may be objected that the estimate of "domestic imports" is too low due to the use of the 0.5 coefficient. Exports were, in any event, sufficiently small, however, so that use of zero would not affect the order of magnitude of the estimate significantly.

2. The choice of year is important, as the Korean economy was increasingly open as time progressed. Choice of an earlier year would suggest smaller premiums than the estimates based on 1970. Conversely, choice of a later year would raise the estimated premiums.

3. Westphal and Kim, "Industrial Policy and Development" (pp. 3–59), estimate the import demand elasticity to be in the range between 1.1 and 2.7 in absolute value.

4. The relevant elasticities are those taking into account both demand and supply changes resulting from income growth.

5. Large as these numbers are, they may not be unreasonable. Estimates for India and Turkey suggest similar orders of magnitude, and there is some basis for believing that the Korean exchange rate of the 1950s may have been even more overvalued than the Indian or Turkish rate. See Anne O. Krueger, "Political Economy of the Rent-Seeking Society," *American Economic Review*, 1974.

6. See, for example, Cole and Lyman, pp. 156 ff.

7. Obviously, with the rapid growth of GNP after 1961, the increased size of the domestic market would have provided an offset to the impact of increased competition from abroad.

8. BOK, *Economic Statistics Yearbook, 1965*, pp. 172–175.

9. Even if output growth did resume, it may have originated from firms other than those established in the 1950s.

10. There were also unfortunate consequences for macroeconomic policy, which are considered in Chapter 6.

11. Efforts were repeatedly made to encourage use of domestically produced intermediate goods. For example, a "local LIC" system was established in 1965 under which producers of intermediate goods used for export were extended many of the same privileges as exporters. See Frank, Kim, and Westphal, p. 51. That imported inputs remained cheaper than domestic ones is evidenced by the fact that the wastage allowance was regarded as an export incentive.

12. Borrowers typically were able to finance about 70% of their projects through these loans, and resorted to the curb market for the remainder. In general, curb-market rates exceeded 35%. Hong estimates that the average rate of return on capital was also well over 30%.

13. This happened because short-term export credit loans were eligible for unlimited rediscount by the Bank of Korea.

14. Wontack Hong, "Trade, Distortions and Employment," p. 129.

15. See the discussion of Rhee and Westphal's results under export policy below.

16. For a careful statement of the theory of effective protection, see W. M. Corden, *The Theory of Protection* (Oxford, 1971).

17. To be sure, if the industry subject to higher effective protection also has a greater cost disadvantage, a higher ERP may not be associated with more resources being pulled into an industry.

18. See Westphal and Kim, "Industrial Policy and Development," Appendix Table 2.

19. Wontack Hong, in correspondence, has suggested that the different labor coefficients in primary industries in the two sets of estimates are the result of different treatment of agriculture. The Westphal-Kim data are numbers of employed persons in agriculture, while the Hong data are in man-years of labor.

20. See the discussion of employment in Chapter 6.

21. Westphal and Kim, "Industrial Policy and Development," pp. 4–5.

22. Yung W. Rhee and Larry E. Westphal, "A Micro, Econometric Investigation of the Impact of Industrial Policy on Technology Choice," paper presented at the Econometric Society Meetings, Atlantic City, September 16–18, 1976. The article was published in the *Journal of Development Studies,* September 1977, but the quotation cited here was omitted from the published version. Westphal has stated in correspondence, however, that the authors still agree with it.

23. Ibid., p. 47, in the mimeo version.

24. See also the discussion of the macroeconomics of the labor market in Chapter 6.

25. See Chapter 6 for estimates of the contribution of foreign loans to foreign saving, total saving, and economic growth.

26. This is over and above the misallocation resulting from non-optimal industries or techniques. The distinction is that optimal foreign borrowing takes place when the real rate of return equals the interest rate. When the real return is below the real interest rate, as must have happened, there is a net loss from the country to foreigners as the real cost of borrowing exceeds the real return. If that happens domestically, the result is simply a transfer from one part of society to another.

27. Frank, Kim, and Westphal, p. 116–117.

SIX   *Macroeconomic Effects of Trade and Aid*

1. Data on savings are available only in current prices. The wŏn value of net transfers is therefore not indicative of the absolute level of the real resource flow.

2. It should be recalled that, by 1970, most aid took the form of concessional loans, so that "net transfers" can no longer be identified solely with aid flows, and aid was not only in net transfers. A consistent framework for linking aid to net savings is not available except via net transfers. The problem is, in large part, the difficulty of converting dollar flows of aid into wŏn equivalents.

3. Cole and Lyman, p. 170.

4. See Table 44 for estimates of the relative importance of aid in borrowing in the latter part of the 1960s.

5. See Frank, Kim, and Westphal, pp. 107–108.

6. Another avenue by which exports may have contributed is to the extent there were economies to scale in individual manufacturing sectors. Since exporting permitted greater scale in some sectors, more economies of scale were exploitable. The only available estimate is that of Chong Nam. He estimated production functions and found that, for the 1966–1968 period, about 18% of the growth in manufacturing output could be accounted for by economies to scale in individual manufacturing sectors. See his "Economies of Scale and Production Functions in South Korean Manufacturing," Ph.D. dissertation, University of Minnesota, 1975.

7. In the belief that intangibles of export promotion may well be of great significance in determining growth performance, for other purposes I attempted to treat the time series observations of the ten countries included in the National Bureau of Economic Research project on Foreign Trade Regimes and Economic Development together, in effect, "pooling time series and cross section." The separate rate of growth of each country's real GNP was estimated as a function of time, and then a common estimator was obtained for the effect of varying the rate of growth of export earnings. The results of those estimates, which are subject to numerous qualifications, implied that a 1% increase in the rate of export growth leads, on average, to more rapid GNP growth by one-tenth of 1%. If that estimate is then used on Korea alone, it implies that her 40% average rate of growth of exports from 1960 to 1973 accounted for about 4 percentage points of real GNP growth annually. If that estimate is combined with the rough calculations of the contribution of foreign savings given above, the "guesstimate" would be that about 8 percentage points of

growth in the 1960s and early 1970s are accounted for by the trade and payments regime and capital flows. See Anne O. Krueger, *Foreign Trade Regimes and Economic Development: Liberalization Attempts and Consequences*, (National Bureau of Economic Research, 1978).

8. Westphal and Kim, "Industrial Policy and Development," p. 109.

9. Of course, part of domestic demand expansion may have resulted from the multiplier effects of export growth. If, however, demand management would anyway have been satisfactory, one cannot attribute multiplier effects to export growth.

10. For full details, see Frank, Kim, and Westphal, Chapter 9.

11. Ibid., p. 184.

12. Ibid., p. 220.

13. Frank, Kim, and Westphal give average real monthly earnings (at 1970 prices) as 7,778 wŏn in 1957, rising then to a peak of 8,902 wŏn in 1959. Thereafter, they fluctuated between 7,549 wŏn (in 1964) and 8,540 wŏn (in 1962), with no discernible trend until 1967. In that year, average earnings rose to 9,159 wŏn. Thereafter, the growth of real wages was rapid, averaging 15% per year between 1967 and 1970. *Foreign Trade Regimes*, p. 222.

14. Susumu Watanabe, "Exports and Employment: The Case of the Republic of Korea," *International Labor Review*, (December 1972), pp. 495–526.

15. David C. Cole and Larry E. Westphal, "The Contribution of Exports to Employment in Korea," in Hong and Krueger, *Trade and Development*, pp. 89–102.

16. For a comparison of the two methods, see ibid., pp. 96 ff.

17. The breakdown of employment by industry, however, is quite different.

APPENDIX A   *Definition of Exchange Rate Terms*

1. These concepts were first used systematically in the National Bureau of Economic Research Project on Foreign Trade Regimes and Economic Development. See Krueger, *Foreign Trade Regimes*, for a fuller discussion.

Bibliography

Ban, Sung Hwan (Pan, Sŏng-hwan). *Growth Rates of Korean Agriculture 1968-71.* Seoul, Korea Development Institute Press, 1974.

Bank of Korea. *Annual Economic Review.* Seoul, 1955.

Bank of Korea. *Economic Review.* Seoul, 1955.

Bank of Korea. *Economics Statistics Yearbook, 1953* through *1976.* Seoul. (In both English and Korean).

Bank of Korea. *Monthly Statistical Review.* February 1952.

Bank of Korea, Research Department. *Price Statistics Summary.* Seoul. 1964.

Bloomfield, Arthur I. and John P. Jensen. *Banking Reform and South Korea.* New York, Federal Reserve Bank of New York, March, 1951.

Chōsen Sōtokufu (Japanese Government General in Korea). *Chōsen Sōtokufu tōkei nenpō.* (Annual statistical report of the office of the Governor General of Korea). Seoul.

Cole, David C. and Princeton N. Lyman. *Korean Development, The Interplay of Politics and Economics.* Cambridge, Harvard University Press, 1971.

Cole, David C. and Larry E. Westphal. "The Contribution of Exports to Employment in Korea," in Wontack Hong (Wŏn-t'aek Hong) and Anne O. Krueger, *Trade and Development in Korea.* Seoul, Korea Development Institute Press, 1975.

Corden, W. M. *The Theory of Protection.* Oxford, Clarendon Press, 1971.

Economic Planning Board. "Foreign Capital Inducement and Investment Policy." Mimeographed. Seoul, 1975.

Frank, Charles R., Kwang Suk Kim (Kwang-sŏk Kim) and Larry E. Westphal. *Foreign Trade Regimes and Economic Development: South Korea.* New York, National Bureau of Economic Research, 1975.

Hasan, Parvez. *Korea, Problems and Issues in a Rapidly Growing Economy.* Baltimore, Johns Hopkins University Press, 1976.

Henderson, Gregory. *Korea, The Politics of the Vortex.* Cambridge, Harvard University Press, 1968.

Hong, Wontack (Hong, Wŏn-t'aek). *Factor Supply and Factor Intensity of Trade in Korea.* Seoul, Korea Development Institute Press, 1975.

——. "Trade, Distortions and Employment Growth in Korea," and Statistical Appendix thereto. Mimeographed. Seoul, Korea Development Institute Press, 1977.

International Monetary Fund. *Annual Report, 1973, 1975.* Washington, D.C.

International Monetary Fund. *Annual Report on Exchange Restrictions, 1956-1975.* Washington, D.C.

International Monetary Fund. *International Financial Statistics,* various issues. Washington, D.C., 1976.

Johnson, E.A.G. *American Imperialism in the Image of Peer Gynt: Memoirs of a Professor-Bureaucrat.* Minneapolis, University of Minnesota Press, 1971.

Jones, Leroy. *Public Enterprises and Economic Development: The Case of Korea.* Seoul, Korea Development Institute Press, 1975.

Kim, Kwang Suk (Kim, Kwang-sŏk). "Outward-Looking Industrialization Strategy: The Case of Korea," in Wontack Hong (Wŏn-t'aek Hong) and Anne O. Krueger, eds., *Trade and Development in Korea.* Seoul, Korea Development Institute Press, 1975.

—— and Michael Roemer. *Growth and Structural Transformation,* Studies in the Modernization of the Republic of Korea: 1945-1975. Cambridge, Council on East Asian Studies, Harvard University, 1978.

Koh, Harold. "The Early History of U.S. Economic Assistance to the Republic of Korea, 1955-63." Typed. 1975.

Krueger, Anne O. "Political Economy of the Rent-Seeking Society," *American Economic Review,* March, 1974.

——. *Foreign Trade Regimes and Economic Development: Liberalization Attempts and Consequences.* National Bureau of Economic Research. Cambridge, Ballinger Press, 1978.

Lewis, John P. "Reconstruction and Development in South Korea." Pamphlet 94. Washington, D.C., National Planning Association. December, 1955.

Lyons, Gene M. *Military Policy and Economic Aid: The Korean Case, 1950-1953.* Colombus, Ohio State University Press, 1961.

Mitchell, Clyde. "Land Reform in Asia, a Case Study." Pamphlet 78. Washington, D.C., National Planning Association, 1952.

Nam, Chong. "Economies of Scale and Production Functions in South Korean Manufacturing." Ph.D. dissertation, University of Minnesota, 1975.

Ouchi, H. ed. *Nihon keizai tōkeishū* (Japanese economic statistics). Tokyo, Nihon Tōkei Kenkyūjo, 1958.

Reeve, W. D. *The Republic of Korea: A Political and Economic Study.* London, Oxford University Press, 1963.

Rhee, Yung W. and Larry E. Westphal. "A Micro, Econometric Investigation of the Impact of Industrial Policy on Technology Choice," *Journal of Development Studies.* London, Frank Cass & Co., Ltd., September 1977.

Suh, Suk Tai. "Import Substitution and Economic Development in Korea." Seoul, Korea Development Institute Press, 1975.

——. "Foreign Assistance in Modernization of Korea." Seoul, Korea Development Institute Press, 1975.

——. *Statistical Report on Foreign Assistance and Loans to Korea (1945-75).* Monograph 7602. Seoul, Korea Development Institute Press, 1976.

——. "Revision of Tariff Rates and the Introduction of Flexible Tariff Systems." Seoul, Korea Development Institute, July 5, 1976.

Tewksbury, Donald G. *Source Materials on Korean Politics and Ideologies.* New York, Institute of Pacific Relations, 1950.

United Nations. *Yearbook of International Trade Statistics, 1955, 1956, 1957.*

U.S. Department of State, Economic Cooperation Administration. "ECA Recovery Program for 1950." Mimeographed. Washington, D.C., June, 1949.

Watanabe, Susumu. "Exports and Employment: The Case of the Republic of Korea," *International Labor Review*, Geneva, December, 1972.

Westphal, Larry E. and Kwang Suk Kim (Kwang-sŏk Kim). "Industrial Policy and Development in Korea." Mimeographed. Washington, D.C.: International Bank for Reconstruction and Development. World Bank. 1974, later revised. World Bank Staff Working Paper No. 263. Washington, D.C., International Bank for Reconstruction and Development, August 1977.

Index

\

Harvard East Asian Monographs

1. Liang Fang-chung, *The Single-Whip Method of Taxation in China*
2. Harold C. Hinton, *The Grain Tribute System of China, 1845–1911*
3. Ellsworth C. Carlson, *The Kaiping Mines, 1877–1912*
4. Chao Kuo-chün, *Agrarian Policies of Mainland China: A Documentary Study, 1949–1956*
5. Edgar Snow, *Random Notes on Red China, 1936–1945*
6. Edwin George Beal, Jr., *The Origin of Likin, 1835–1864*
7. Chao Kuo-chün, *Economic Planning and Organization in Mainland China: A Documentary Study, 1949–1957*
8. John K. Fairbank, *Ch'ing Documents: An Introductory Syllabus*
9. Helen Yin and Yi-chang Yin, *Economic Statistics of Mainland China, 1949–1957*
10. Wolfgang Franke, *The Reform and Abolition of the Traditional Chinese Examination System*
11. Albert Feuerwerker and S. Cheng, *Chinese Communist Studies of Modern Chinese History*
12. C. John Stanley, *Late Ch'ing Finance: Hu Kuang-yung as an Innovator*
13. S. M. Meng, *The Tsungli Yamen: Its Organization and Functions*
14. Ssu-yü Teng, *Historiography of the Taiping Rebellion*
15. Chun-Jo Liu, *Controversies in Modern Chinese Intellectual History: An Analytic Bibliography of Periodical Articles, Mainly of the May Fourth and Post-May Fourth Era*
16. Edward J. M. Rhoads, *The Chinese Red Army, 1927–1963: An Annotated Bibliography*
17. Andrew J. Nathan, *A History of the China International Famine Relief Commission*
18. Frank H. H. King (ed.) and Prescott Clarke, *A Research Guide to China-Coast Newspapers, 1822–1911*
19. Ellis Joffe, *Party and Army: Professionalism and Political Control in the Chinese Officer Corps, 1949–1964*
20. Toshio G. Tsukahira, *Feudal Control in Tokugawa Japan: The Sankin Kōtai System*

46. W. P. J. Hall, *A Bibliographical Guide to Japanese Research on the Chinese Economy, 1958–1970*

47. Jack J. Gerson, *Horatio Nelson Lay and Sino-British Relations, 1854–1864*

48. Paul Richard Bohr, *Famine and the Missionary: Timothy Richard as Relief Administrator and Advocate of National Reform*

49. Endymion Wilkinson, *The History of Imperial China: A Research Guide*

50. Britten Dean, *China and Great Britain: The Diplomacy of Commerical Relations, 1860–1864*

51. Ellsworth C. Carlson, *The Foochow Missionaries, 1847–1880*

52. Yeh-chien Wang, *An Estimate of the Land-Tax Collection in China, 1753 and 1908*

53. Richard M. Pfeffer, *Understanding Business Contracts in China, 1949–1963*

54. Han-sheng Chuan and Richard Kraus, *Mid-Ch'ing Rice Markets and Trade, An Essay in Price History*

55. Ranbir Vohra, *Lao She and the Chinese Revolution*

56. Liang-lin Hsiao, *China's Foreign Trade Statistics, 1864–1949*

57. Lee-hsia Hsu Ting, *Government Control of the Press in Modern China, 1900–1949*

58. Edward W. Wagner, *The Literati Purges: Political Conflict in Early Yi Korea*

59. Joungwon A. Kim, *Divided Korea: The Politics of Development, 1945–1972*

60. Noriko Kamachi, John K. Fairbank, and Chūzō Ichiko, *Japanese Studies of Modern China Since 1953: A Bibliographical Guide to Historical and Social-Science Research on the Nineteenth and Twentieth Centuries, Supplementary Volume for 1953–1969*

61. Donald A. Gibbs and Yun-chen Li, *A Bibliography of Studies and Translations of Modern Chinese Literature, 1918–1942*

62. Robert H. Silin, *Leadership and Values: The Organization of Large-Scale Taiwanese Enterprises*

63. David Pong, *A Critical Guide to the Kwangtung Provincial Archives Deposited at the Public Record Office of London*

64. Fred W. Drake, *China Charts the World: Hsu Chi-yü and His Geography of 1848*

65. William A. Brown and Urgunge Onon, translators and annotators, *History of the Mongolian People's Republic*

66. Edward L. Farmer, *Early Ming Government: The Evolution of Dual Capitals*

67. Ralph C. Croizier, *Koxinga and Chinese Nationalism: History, Myth, and the Hero*

68. William J. Tyler, tr., *The Psychological World of Natsumi Sōseki*, by Doi Takeo

STUDIES IN THE MODERNIZATION OF THE REPUBLIC OF KOREA: 1945–1975

91. Leroy P. Jones and Il SaKong, *Government, Business, and Entrepreneurship in Economic Development: The Korean Case*

92. Edward S. Mason, Dwight H. Perkins, Kwang Suk Kim, David C. Cole, Mahn Je Kim, et al., *The Economic and Social Modernization of the Republic of Korea*